AF889173

Risk Management Framework

Auditors' Survival Guide

Dennis Adjepong MD. MBA

ISBN Paperback: 978-1-964467-13-9

ISBN Hardcover: 978-1-964467-14-6

LCCN: 2024916750

Dedication

This book is dedicated to all IT auditors worldwide, ISSOs, and assessors who pursue auditing using Risk management methodology.

Table of Contents

CNTL NO.	CONTROL NAME	PRIORITY	INITIAL CONTROL BASELINES		
			LOW	MOD	HIGH
			Access Control		
AC-1	Access Control Policy and Procedures	P1	AC-1	AC-1	AC-1
AC-2	Account Management	P1	AC-2	AC-2 (1) (2) (3) (4)	AC-2 (1) (2) (3) (4) (5) (12) (13)
AC-3	Access Enforcement	P1	AC-3	AC-3	AC-3
AC-4	Information Flow Enforcement	P1	Not Selected	AC-4	AC-4
AC-5	Separation of Duties	P1	Not Selected	AC-5	AC-5
AC-6	Least Privilege	P1	Not Selected	AC-6 (1) (2) (5) (9) (10)	AC-6 (1) (2) (3) (5) (9) (10)
AC-7	Unsuccessful Logon Attempts	P2	AC-7	AC-7	AC-7
AC-8	System Use Notification	P1	AC-8	AC-8	AC-8
AC-9	Previous Logon (Access) Notification	P0	Not Selected	Not Selected	Not Selected
AC-10	Concurrent Session Control	P2	Not Selected	Not Selected	AC-10
AC-11	Session Lock	P3	Not Selected	AC-11 (1)	AC-11 (1)
AC-12	**Withdrawn**	---	---	---	---
AC-13	**Withdrawn**	---	---	---	---
AC-14	Permitted Actions without Identification or Authentication	P1	AC-14	AC-14	AC-14
AC-15	**Withdrawn**	---	---	---	---
AC-16	Security Attributes	P0	Not Selected	Not Selected	Not Selected
AC-17	Remote Access	P1	AC-17	AC-17 (1) (2) (3) (4)	AC-17 (1) (2) (3) (4)
AC-18	Wireless Access	P1	AC-18	AC-18 (1)	AC-18 (1) (4) (5)
AC-19	Access Control for Mobile Devices	P1	AC-19	AC-19 (5)	AC-19 (5)
AC-20	Use of External Information Systems	P1	AC-20	AC-20 (1) (2)	AC-20 (1) (2)
AC-21	Information Sharing	P0	Not Selected	AC-21	AC-21
AC-22	Publicly Accessible Content	P2	AC-22	AC-22	AC-22
AC-23	Data Mining Protection	P0	Not Selected	Not Selected	Not Selected
AC-24	Access Control Decisions	P0	Not Selected	Not Selected	Not Selected

CNTL NO.	CONTROL NAME	PRIORITY	INITIAL CONTROL BASELINES		
			LOW	MOD	HIGH
Access Control					
AC-25	Reference Monitor	P0	Not Selected	Not Selected	Not Selected
Awareness and Training					
AT-1	Security Awareness and Training Policy and Procedures	P1	AT-1	AT-1	AT-1
AT-2	Security Awareness Training	P1	AT-2	AT-2 (2)	AT-2 (2)
AT-3	Role-Based Security Training	P1	AT-3	AT-3	AT-3
AT-4	Security Training Records	P3	AT-4	AT-4	AT-4
AT-5	**Withdrawn**	---	---	---	---
Audit and Accountability					
AU-1	Audit and Accountability Policy and Procedures	P1	AU-1	AU-1	AU-1
AU-2	Audit Events	P1	AU-2	AU-2 (3)	AU-2 (3)
AU-3	Content of Audit Records	P1	AU-3	AU-3 (1)	AU-3 (1) (2)
AU-4	Audit Storage Capacity	P1	AU-4	AU-4	AU-4
AU-5	Response to Audit Processing Failures	P1	AU-5	AU-5	AU-5 (1) (2)
AU-6	Audit Review, Analysis, and Reporting	P1	AU-6	AU-6 (1) (3)	AU-6 (1) (3) (5) (6)
AU-7	Audit Reduction and Report Generation	P2	Not Selected	AU-7 (1)	AU-7 (1)
AU-8	Time Stamps	P1	AU-8	AU-8 (1)	AU-8 (1)
AU-9	Protection of Audit Information	P1	AU-9	AU-9 (4)	AU-9 (2) (3) (4)
AU-10	Non-repudiation	P1	Not Selected	Not Selected	Not Selected
AU-11	Audit Record Retention	P3	AU-11	AU-11	AU-11
AU-12	Audit Generation	P1	AU-12	AU-12	AU-12 (1) (3)
AU-13	Monitoring for Information Disclosure	P0	Not Selected	Not Selected	Not Selected
AU-14	Session Audit	P0	Not Selected	Not Selected	Not Selected
AU-15	Alternate Audit Capability	P0	Not Selected	Not Selected	Not Selected
AU-16	Cross-Organizational Auditing	P0	Not Selected	Not Selected	Not Selected
Security Assessment and Authorization					
CA-1	Security Assessment and Authorization Policies and Procedures	P1	CA-1	CA-1	CA-1
CA-2	Security Assessments	P2	CA-2	CA-2 (1)	CA-2 (1) (2)
CA-3	System Interconnections	P1	CA-3	CA-3 (5)	CA-3 (5)
CA-4	**Withdrawn**	---	---	---	---
CA-5	Plan of Action and Milestones	P3	CA-5	CA-5	CA-5
CA-6	Security Authorization	P3	CA-6	CA-6	CA-6
CA-7	Continuous Monitoring	P3	CA-7	CA-7 (1)	CA-7 (1)
CA-8	Penetration Testing	P1	Not Selected	Not Selected	CA-8
CA-9	Internal System Connections	P1	CA-9	CA-9	CA-9
Configuration Management					
CM-1	Configuration Management Policy and Procedures	P1	CM-1	CM-1	CM-1
CM-2	Baseline Configuration	P1	CM-2	CM-2 (1) (3) (7)	CM-2 (1) (2) (3) (7)
CM-3	Configuration Change Control	P1	Not Selected	CM-3	CM-3 (1)
CM-4	Security Impact Analysis	P2	CM-4	CM-4 (3)	CM-4 (1) (3)

CNTL NO.	CONTROL NAME	PRIORITY	INITIAL CONTROL BASELINES		
			LOW	MOD	HIGH
CM-5	Access Restrictions for Change	P1	Not Selected	CM-5	CM-5 (1) (2) (3)
CM-6	Configuration Settings	P1	CM-6	CM-6	CM-6 (1) (2)
CM-7	Least Functionality	P1	CM-7	CM-7 (1) (2) (4)	CM-7 (1) (2) (5)
CM-8	Information System Component Inventory	P1	CM-8	CM-8 (1) (3) (5)	CM-8 (1) (2) (3) (4) (5)
CM-9	Configuration Management Plan	P1	Not Selected	CM-9	CM-9
CM-10	Software Usage Restrictions	P1	CM-10	CM-10	CM-10
CM-11	User-Installed Software	P1	CM-11	CM-11	CM-11
Contingency Planning					
CP-1	Contingency Planning Policy andProcedures	P1	CP-1	CP-1	CP-1
CP-2	Contingency Plan	P1	CP-2	CP-2 (1) (3) (8)	CP-2 (1) (2) (3) (4) (5) (8)
CP-3	Contingency Training	P2	CP-3	CP-3	CP-3 (1)
CP-4	Contingency Plan Testing	P2	CP-4	CP-4 (1)	CP-4 (1) (2)
CP-5	**Withdrawn**	---	---	---	---
CP-6	Alternate Storage Site	P1	Not Selected	CP-6 (1) (3)	CP-6 (1) (2) (3)
CP-7	Alternate Processing Site	P1	Not Selected	CP-7 (1) (2) (3)	CP-7 (1) (2) (3) (4)
CP-8	Telecommunications Services	P1	Not Selected	CP-8 (1) (2)	CP-8 (1) (2) (3) (4)
CP-9	Information System Backup	P1	CP-9	CP-9 (1)	CP-9 (1) (2) (3) (5)
CP-10	Information System Recovery andReconstitution	P1	CP-10	CP-10 (2)	CP-10 (2) (4)
CP-11	Alternate Communications Protocols	P0	Not Selected	Not Selected	Not Selected
CP-12	Safe Mode	P0	Not Selected	Not Selected	Not Selected
CP-13	Alternative Security Mechanisms	P0	Not Selected	Not Selected	Not Selected
Identification and Authentication					
IA-1	Identification and Authentication Policy and Procedures	P1	IA-1	IA-1	IA-1
IA-2	Identification and Authentication(Organizational Users)	P1	IA-2 (1) (13)	IA-2 (1) (2) (3) (8) (11) (12) (13)	IA-2 (1) (2) (3) (4) (8) (9) (11) (12) (13)
IA-3	Device Identification and Authentication	P1	Not Selected	IA-3	IA-3
IA-4	Identifier Management	P1	IA-4	IA-4	IA-4
IA-5	Authenticator Management	P1	IA-5 (1) (11)	IA-5 (1) (2) (3) (11)	IA-5 (1) (2) (3) (11)
IA-6	Authenticator Feedback	P1	IA-6	IA-6	IA-6
IA-7	Cryptographic Module Authentication	P1	IA-7	IA-7	IA-7
IA-8	Identification and Authentication (Non-Organizational Users)	P1	IA-8 (1) (2) (3) (4)	IA-8 (1) (2) (3) (4)	IA-8 (1) (2) (3) (4)
IA-9	Service Identification and Authentication	P0	Not Selected	Not Selected	Not Selected
IA-10	Adaptive Identification and Authentication	P0	Not Selected	Not Selected	Not Selected
IA-11	Re-authentication	P0	Not Selected	Not Selected	Not Selected
Incident Response					
IR-1	Incident Response Policy and Procedures	P1	IR-1	IR-1	IR-1

CNTL NO.	CONTROL NAME	PRIORITY	INITIAL CONTROL BASELINES		
			LOW	MOD	HIGH
IR-2	Incident Response Training	P2	IR-2	IR-2	IR-2 (1) (2)
IR-3	Incident Response Testing	P2	Not Selected	IR-3 (2)	IR-3 (2)
IR-4	Incident Handling	P1	IR-4	IR-4 (1)	IR-4 (1) (4)
IR-5	Incident Monitoring	P1	IR-5	IR-5	IR-5 (1)
IR-6	Incident Reporting	P1	IR-6	IR-6 (1)	IR-6 (1)
IR-7	Incident Response Assistance	P3	IR-7	IR-7 (1)	IR-7 (1)
IR-8	Incident Response Plan	P1	IR-8	IR-8	IR-8
IR-9	Information Spillage Response	P0	Not Selected	Not Selected	Not Selected
IR-10	Integrated Information Security Cell	P0	Not Selected	Not Selected	Not Selected
Maintenance					
MA-1	System Maintenance Policy and Procedures	P1	MA-1	MA-1	MA-1
MA-2	Controlled Maintenance	P2	MA-2	MA-2	MA-2 (2)
MA-3	Maintenance Tools	P2	Not Selected	MA-3 (1) (2)	MA-3 (1) (2) (3)
MA-4	Nonlocal Maintenance	P1	MA-4	MA-4 (2)	MA-4 (2) (3)
MA-5	Maintenance Personnel	P1	MA-5	MA-5	MA-5 (1)
MA-6	Timely Maintenance	P1	Not Selected	MA-6	MA-6
Media Protection					
MP-1	Media Protection Policy and Procedures	P1	MP-1	MP-1	MP-1
MP-2	Media Access	P1	MP-2	MP-2	MP-2
MP-3	Media Marking	P1	Not Selected	MP-3	MP-3
MP-4	Media Storage	P1	Not Selected	MP-4	MP-4
MP-5	Media Transport	P1	Not Selected	MP-5 (4)	MP-5 (4)
MP-6	Media Sanitization	P1	MP-6	MP-6	MP-6 (1) (2) (3)
MP-7	Media Use	P1	MP-7	MP-7 (1)	MP-7 (1)
MP-8	Media Downgrading	P0	Not Selected	Not Selected	Not Selected
Physical and Environmental Protection					
PE-1	Physical and Environmental ProtectionPolicy and Procedures	P1	PE-1	PE-1	PE-1
PE-2	Physical Access Authorizations	P1	PE-2	PE-2	PE-2
PE-3	Physical Access Control	P1	PE-3	PE-3	PE-3 (1)
PE-4	Access Control for Transmission Medium	P1	Not Selected	PE-4	PE-4
PE-5	Access Control for Output Devices	P1	Not Selected	PE-5	PE-5
PE-6	Monitoring Physical Access	P1	PE-6	PE-6 (1)	PE-6 (1) (4)
PE-7	**Withdrawn**	---	---	---	---
PE-8	Visitor Access Records	P3	PE-8	PE-8	PE-8 (1)
PE-9	Power Equipment and Cabling	P1	Not Selected	PE-9	PE-9
PE-10	Emergency Shutoff	P1	Not Selected	PE-10	PE-10
PE-11	Emergency Power	P1	Not Selected	PE-11	PE-11 (1)
PE-12	Emergency Lighting	P1	PE-12	PE-12	PE-12
PE-13	Fire Protection	P1	PE-13	PE-13 (3)	PE-13 (1) (2) (3)
PE-14	Temperature and Humidity Controls	P1	PE-14	PE-14	PE-14
PE-15	Water Damage Protection	P1	PE-15	PE-15	PE-15 (1)
PE-16	Delivery and Removal	P1	PE-16	PE-16	PE-16

PE-17	Alternate Work Site	P1	Not Selected	PE-17	PE-17
PE-18	Location of Information System Components	P2	Not Selected	Not Selected	PE-18
PE-19	Information Leakage	P0	Not Selected	Not Selected	Not Selected
	Planning				
PL-1	Security Planning Policy and Procedures	P1	PL-1	PL-1	PL-1
PL-2	System Security Plan	P1	PL-2	PL-2 (3)	PL-2 (3)
PL-3	**Withdrawn**	---	---	---	---
PL-4	Rules of Behavior	P1	PL-4	PL-4 (1)	PL-4 (1)
PL-5	**Withdrawn**	---	---	---	---
PL-6	**Withdrawn**	---	---	---	---
PL-7	Security Concept of Operations	P0	Not Selected	Not Selected	Not Selected
PL-8	Information Security Architecture	P0	Not Selected	PL-8	PL-8
PL-9	Central Management	P0	Not Selected	Not Selected	Not Selected
	Personnel Security				
PS-1	Personnel Security Policy and Procedures	P1	PS-1	PS-1	PS-1
PS-2	Position Risk Designation	P1	PS-2	PS-2	PS-2
PS-3	Personnel Screening	P1	PS-3	PS-3	PS-3
PS-4	Personnel Termination	P2	PS-4	PS-4	PS-4 (2)
PS-5	Personnel Transfer	P2	PS-5	PS-5	PS-5
PS-6	Access Agreements	P3	PS-6	PS-6	PS-6
PS-7	Third-Party Personnel Security	P1	PS-7	PS-7	PS-7
PS-8	Personnel Sanctions	P3	PS-8	PS-8	PS-8
	Risk Assessment				
RA-1	Risk Assessment Policy and Procedures	P1	RA-1	RA-1	RA-1
RA-2	Security Categorization	P1	RA-2	RA-2	RA-2
RA-3	Risk Assessment	P1	RA-3	RA-3	RA-3
RA-4	**Withdrawn**	---	---	---	---
RA-5	Vulnerability Scanning	P1	RA-5	RA-5 (1) (2) (5)	RA-5 (1) (2) (4) (5)
RA-6	Technical Surveillance Countermeasures Survey	P0	Not Selected	Not Selected	Not Selected
	System and Services Acquisition				
SA-1	System and Services Acquisition Policy and Procedures	P1	SA-1	SA-1	SA-1
SA-2	Allocation of Resources	P1	SA-2	SA-2	SA-2
SA-3	System Development Life Cycle	P1	SA-3	SA-3	SA-3
SA-4	Acquisition Process	P1	SA-4 (10)	SA-4 (1) (2) (10)	SA-4 (1) (2) (10)
SA-5	Information System Documentation	P2	SA-5	SA-5	SA-5
SA-6	**Withdrawn**	---	---	---	---
SA-7	**Withdrawn**	---	---	---	---
SA-8	Security Engineering Principles	P1	Not Selected	SA-8	SA-8
SA-9	External Information System Services	P1	SA-9	SA-9 (2)	SA-9 (2)
SA-10	Developer Configuration Management	P1	Not Selected	SA-10	SA-10

CNTL NO.	CONTROL NAME	PRIORITY	INITIAL CONTROL BASELINES		
			LOW	MOD	HIGH
SA-11	Developer Security Testing and Evaluation	P2	Not Selected	SA-11	SA-11
SA-12	Supply Chain Protection	P1	Not Selected	Not Selected	SA-12
SA-13	Trustworthiness	P1	Not Selected	Not Selected	Not Selected
SA-14	Criticality Analysis	P0	Not Selected	Not Selected	Not Selected
SA-15	Development Process, Standards, andTools	P2	Not Selected	Not Selected	SA-15
SA-16	Developer-Provided Training	P2	Not Selected	Not Selected	SA-16
SA-17	Developer Security Architecture and Design	P1	Not Selected	Not Selected	SA-17
SA-18	Tamper Resistance and Detection	P0	Not Selected	Not Selected	Not Selected
SA-19	Component Authenticity	P0	Not Selected	Not Selected	Not Selected
SA-20	Customized Development of CriticalComponents	P0	Not Selected	Not Selected	Not Selected
SA-21	Developer Screening	P0	Not Selected	Not Selected	Not Selected
System and Communications Protection					
SC-1	System and Communications ProtectionPolicy and Procedures	P1	SC-1	SC-1	SC-1
SC-2	Application Partitioning	P1	Not Selected	SC-2	SC-2
SC-3	Security Function Isolation	P1	Not Selected	Not Selected	SC-3
SC-4	Information in Shared Resources	P1	Not Selected	SC-4	SC-4
SC-5	Denial of Service Protection	P1	SC-5	SC-5	SC-5
SC-6	Resource Availability	P0	Not Selected	Not Selected	Not Selected
SC-7	Boundary Protection	P1	SC-7	SC-7 (3) (4) (5) (7)	SC-7 (3) (4) (5) (7) (8) (21)
SC-8	Transmission Confidentiality and Integrity	P1	Not Selected	SC-8 (1)	SC-8 (1)
SC-9	**Withdrawn**	---	---	---	---
SC-10	Network Disconnect	P2	Not Selected	SC-10	SC-10
SC-11	Trusted Path	P0	Not Selected	Not Selected	Not Selected
SC-12	Cryptographic Key Establishment andManagement	P1	SC-12	SC-12	SC-12 (1)
SC-13	Cryptographic Protection	P1	SC-13	SC-13	SC-13
SC-14	**Withdrawn**	---	---	---	---
SC-15	Collaborative Computing Devices	P1	SC-15	SC-15	SC-15
SC-16	Transmission of Security Attributes	P0	Not Selected	Not Selected	Not Selected
SC-17	Public Key Infrastructure Certificates	P1	Not Selected	SC-17	SC-17
SC-18	Mobile Code	P1	Not Selected	SC-18	SC-18
SC-19	Voice Over Internet Protocol	P1	Not Selected	SC-19	SC-19
SC-20	Secure Name /Address Resolution Service(Authoritative Source)	P1	SC-20	SC-20	SC-20
SC-21	Secure Name /Address Resolution Service(Recursive or Caching Resolver)	P1	SC-21	SC-21	SC-21
SC-22	Architecture and Provisioning for Name/Address Resolution Service	P1	SC-22	SC-22	SC-22
SC-23	Session Authenticity	P1	Not Selected	SC-23	SC-23
SC-24	Fail in Known State	P1	Not Selected	Not Selected	SC-24
SC-25	Thin Nodes	P0	Not Selected	Not Selected	Not Selected
SC-26	Honeypots	P0	Not Selected	Not Selected	Not Selected

CNTL NO.	CONTROL NAME	PRIORITY	INITIAL CONTROL BASELINES		
			LOW	MOD	HIGH
SC-27	Platform-Independent Applications	P0	Not Selected	Not Selected	Not Selected
SC-28	Protection of Information at Rest	P1	Not Selected	SC-28	SC-28
SC-29	Heterogeneity	P0	Not Selected	Not Selected	Not Selected
SC-30	Concealment and Misdirection	P0	Not Selected	Not Selected	Not Selected
SC-31	Covert Channel Analysis	P0	Not Selected	Not Selected	Not Selected
SC-32	Information System Partitioning	P1	Not Selected	Not Selected	Not Selected
SC-33	**Withdrawn**	---	---	---	---
SC-34	Non-Modifiable Executable Programs	P0	Not Selected	Not Selected	Not Selected
SC-35	Honeyclients	P0	Not Selected	Not Selected	Not Selected
SC-36	Distributed Processing and Storage	P0	Not Selected	Not Selected	Not Selected
SC-37	Out-of-Band Channels	P0	Not Selected	Not Selected	Not Selected
SC-38	Operations Security	P0	Not Selected	Not Selected	Not Selected
SC-39	Process Isolation	P1	SC-39	SC-39	SC-39
SC-40	Wireless Link Protection	P0	Not Selected	Not Selected	Not Selected
SC-41	Port and I/O Device Access	P0	Not Selected	Not Selected	Not Selected
SC-42	Sensor Data	P0	Not Selected	Not Selected	Not Selected
SC-43	Usage Restrictions	P0	Not Selected	Not Selected	Not Selected
SC-44	Detonation Chambers	P0	Not Selected	Not Selected	Not Selected
System and Information Integrity					
SI-1	System and Information Integrity Policy and Procedures	P1	SI-1	SI-1	SI-1
SI-2	Flaw Remediation	P1	SI-2	SI-2 (2)	SI-2 (1) (2)
SI-3	Malicious Code Protection	P1	SI-3	SI-3 (1) (2)	SI-3 (1) (2)
SI-4	Information System Monitoring	P1	SI-4	SI-4 (2) (4) (5)	SI-4 (2) (4) (5)
SI-5	Security Alerts, Advisories, and Directives	P1	SI-5	SI-5	SI-5 (1)
SI-6	Security Function Verification	P1	Not Selected	Not Selected	SI-6
SI-7	Software, Firmware, and Information Integrity	P1	Not Selected	SI-7 (1) (7)	SI-7 (1) (2) (5) (7) (14)
SI-8	Spam Protection	P1	Not Selected	SI-8 (1) (2)	SI-8 (1) (2)
SI-9	**Withdrawn**	---	---	---	---
SI-10	Information Input Validation	P1	Not Selected	SI-10	SI-10
SI-11	Error Handling	P2	Not Selected	SI-11	SI-11
SI-12	Information Handling and Retention	P2	SI-12	SI-12	SI-12
SI-13	Predictable Failure Prevention	P0	Not Selected	Not Selected	Not Selected
SI-14	Non-Persistence	P0	Not Selected	Not Selected	Not Selected
SI-15	Information Output Filtering	P0	Not Selected	Not Selected	Not Selected
SI-16	Memory Protection	P1	Not Selected	SI-16	SI-16
Program Management					
PM-1	Information Security Program Plan	P1			
PM-2	Senior Information Security Officer	P1			
PM-3	Information Security Resources	P1	**Deployed organization-wide. Supporting information security program. Not associated with security control baselines.**		
PM-4	Plan of Action and Milestones Process	P1			
PM-5	Information System Inventory	P1			

Chapter 1

The Assessments I've Done

I have been performing cyber security compliance since 2000 and conducting security assessments since the 1990s. My security career began with a poor Armed Services Aptitude Battery (ASVAB) test score, which led to a post in the Security Forces. Despite my desire to pursue aviation, electronics, or thermonuclear physics, I was assigned to guard those with the necessary skills.

I despised my time in the security forces. It was harsh and terrible work, like a wolf pack that devoured itself. The regular Air Force saw us as inferior due to their perceived intelligence. When we worked with the Army, they said we had it easy, whereas the Marines made their judgments. I disliked the Security Forces, but it taught me that no one can categorize what I like or dislike because there is a task to be done, and failure can result in fatalities. The decision is: "Do you want people to die?" Instead, the question is, "Do you want to do your damn job?"

Security Forces taught me the basics of security: to secure assets, identify potential vulnerabilities, perform assessments, and implement security solutions. This applies to all aspects of security, including guarding C-130 airplanes, producing safe code, and securing networks. Security Forces must undertake regular evaluations.

After successfully retaking the military test, I was able to pursue a career in information technology as a 3COX1 computer operator. My first career goal was to be a network engineer, but I was instead assigned to certification and accreditation, a form of governance, risk, and compliance (GRC).

No one wants to undertake this job in a guarded facility with no windows. I asked everyone how to accomplish it, but no one could help me. I read the information security policy, which referenced several papers. I read the federal statutes that the policy allowed me to. If a directive from the Department of Defense was referenced, I would read it. If additional regulations and procedures were cited, I would also read them. It was like decoding a riddle. Eventually, it made sense. The training was similar to physical security methods taught in Security Forces, but without firearms, conflict, or hostage situations.

Assessments, like physical security, are crucial for ensuring security compliance. Self-assessment and third-party evaluations were undertaken regularly.

Certification and accreditation have been phased out for risk management, authorization, and evaluations. The previous certification and accreditation process resulted in lengthy policy and procedure documents that were rarely reviewed. I'm now focused on ensuring security compliance for both the public and private sectors.

In 2022, terrorist assaults have escalated, but cyber-attacks are much more concerning. Organizations globally face frequent cyber security threats, making the environment increasingly hostile. Security evaluations are more crucial than they have ever been.

Assessments of Panama

In 1989, an activist group infiltrated the Pacific Swan, a ship heading for the Panama Canal. The ship transported 70 tons of radioactive waste from France to Japan via the Panama Canal, a man-made waterway that connects the Atlantic and Pacific oceans.

Activists oppose governments moving nuclear waste across the sea. However, climbing on board and hanging onto a ship's mast is unlikely to prevent this. Greenpeace's representative, Carlos Bravo, stated that they did not want to block the ship's transit through the canal but rather to demonstrate their opposition to nuclear shipment by direct action.

As security experts, we must examine the dire consequences of this. It may have gone differently. Imagine a criminal group, renegade regime, or terror cell gaining entire control over the world's most important waterway. They might restrict international trade. Consider demolishing a portion of the locks to disrupt commerce. The Panama Canal connects the Atlantic and Pacific seas, with over 13,000 ships crossing it annually.

In 1999, the US military returned the Panama Canal to the Republic of Panama, raising security concerns. The suspicions were legitimate due to the Revolutionary Armed Forces of Columbia (FARC) guerillas' frequent incursions into Panamanian territory. The National Liberation Army in Columbia attempted to get attention and funding for their cause by hijacking and conducting massive kidnappings.

Although these parties weren't thinking about the canal, it was a vulnerable target with no defense. All it would take was a little ingenuity and a strategically positioned shoulder-mounted firearm. Following decades of reviews by commercial and public groups, the Panama Canal has received a significant security boost.

In 2004, the Panama Canal developed increased security measures, meeting the International Ship and Port Facility Security (ISPS) Code criteria and receiving fulfillment certification from the American Bureau of Shipping (ABS) Consulting. The Panama Canal Authority (ACP) pursued certification. The ACP was one of the first marine organizations on the continent to get ISPS accreditation. The ISPS code requires the Canal to strengthen its systems for detecting and deterring security risks in port facilities and ships, ensuring a unified framework. The Panama Canal aims to exchange security information, build procedures for assessing security, and enforce safety measures.

Every industry with valuable assets and information requires an evaluation procedure. Neglecting a rigorous evaluation process might leave organizations vulnerable to attackers that exploit flaws. Assessments cannot prevent assaults but can mitigate their damage.

What is an Assessment?

A security control assessment checks and assesses information systems, facilities, and procedures for security characteristics. To better understand what evaluations are, you should learn more about their source. The assessment's context is everything.

Security evaluations rely on industry, private, and federal requirements. The Sarbanes-Oxley Act of 2002, sometimes known as SOX compliance, requires banks and other big financial institutions to conduct audits on monetary transactions and information security. Retail establishments that collect credit card information must follow the Payment Card Industry Data Security Standard (PCI/DSS) and related examinations. Verizon, IBM, and other private corporations provide risk and compliance evaluations for small, medium, and large organizations. Governments also have security rules that must be followed. The US is not the only country with security standards, such as the NIST 800-37 risk management framework that incorporates assessments.

Almost every country has its own set of standards. An examination ensures that security elements are operational and effective. The procedure for assessing security controls varies for each company.

The rules and regulations allow for flexibility in execution, avoiding a "one size fits all" approach. We provide a comprehensive overview of federal legislation to help you navigate any company adopting them.

This book focuses on how the NIST 800 implements assessments. The procedure is detailed in NIST 800-37, 800-53, 800-53A, and 800-115:

- **NIST 800-37:** Risk Management Framework for Information Systems and Organizations: A System Life Cycle Approach to Security and Privacy.
- **NIST 800-53:** Security and Privacy Controls for Information Systems and Organizations.
- **NIST 800-53A:** Assessing Security and Privacy Controls in Information Systems and Organizations.
- **NIST 800-115:** Technical Guide to Information Security Testing and Assessment.
- Understanding how NIST 800 controls are assessed is essential for understanding security compliance frameworks, whether commercial, private, or government.

Quick Overview of the NIST 800-37 Process

Prioritize understanding of the NIST 800 process and controls before delving into evaluation techniques. However, here is a short high-level summary of how the NIST 800-37 RMF operates.

The NIST 800 risk management framework consists of seven steps. This technique seeks to minimize damage to assets after a disaster. Disasters can take many forms, such as earthquakes, server crashes, or attacks by former Soviet Union countries seeking importance.

This step is completed for each significant system inside an "authorization boundary." The phrase "authorization boundary" refers to the fact that the same employer manages many networked systems. Each main group of networked systems follows the NIST 800-37 process:

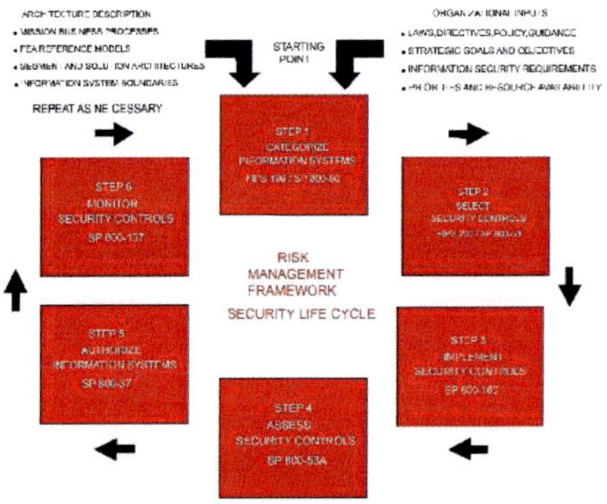

- **Prepare** by coordinating with

stakeholders and gathering data to begin the system's procedure.

- **Categorize:** Determine the security classification of the collection of systems. Categories include low, moderate, and high.
- **Choose:** Decide which controls will be utilized based on the category and systems.
- **Install:** Configure or add security controls to secure the system.
- **Assess:** Analyze security controls to ensure successful implementation.
- **Authorize:** Upper management acknowledges the system's remaining risks in writing.
- **Monitor:** for changes in the system, threats, and environment.

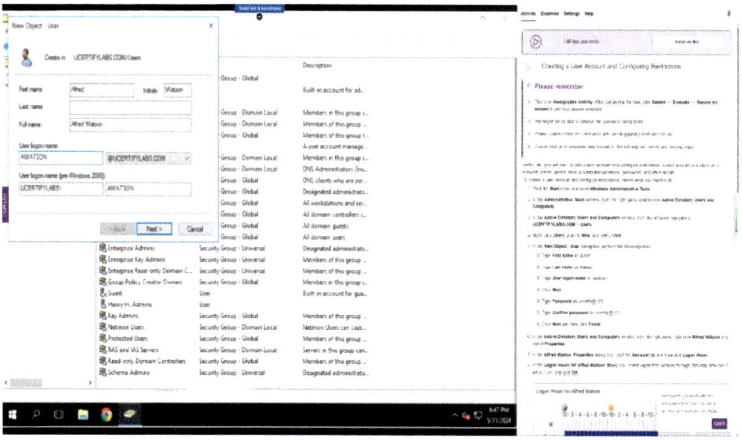

Chapter 2

Role of a Security Controls Assessor

The Cybersecurity and Infrastructure Security Agency (CISA)

A Security Controls Assessor (SCA) conducts independent, comprehensive assessments of the management, operational, and technical security controls and control enhancements employed within or inherited by an information technology (IT) system. The

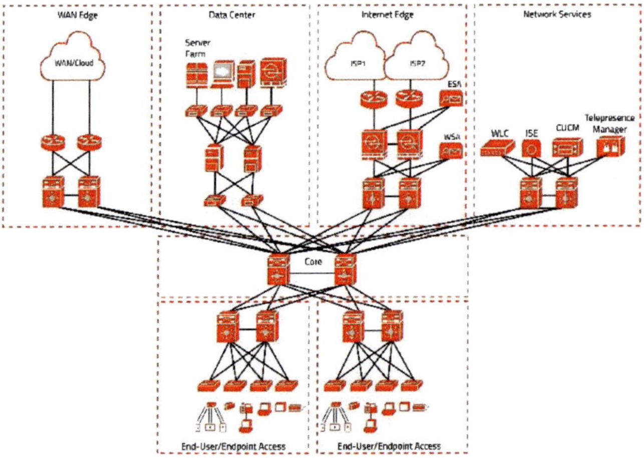

goal is to determine the overall effectiveness of the controls, as defined in NIST SP 800-37. An SCA evaluates the effective implementation of security features on an organization's computers, networks, and procedures.

This role might be an assessor, validator, certifying authority, information assurance compliance analyst, auditor, or SCA (my favourite). Some government agencies use terms like SCA-V or SCA-O. For SCA-V, the "V" stands for the validator. These jobs examine risks before granting or renewing a system license to operate. SCA-O often refers to the "organization" responsible for performing assessments.

We will not include all types of employment in each business, as each organization has its term for the same role. We will focus on essential information, as definitions, names, and acronyms

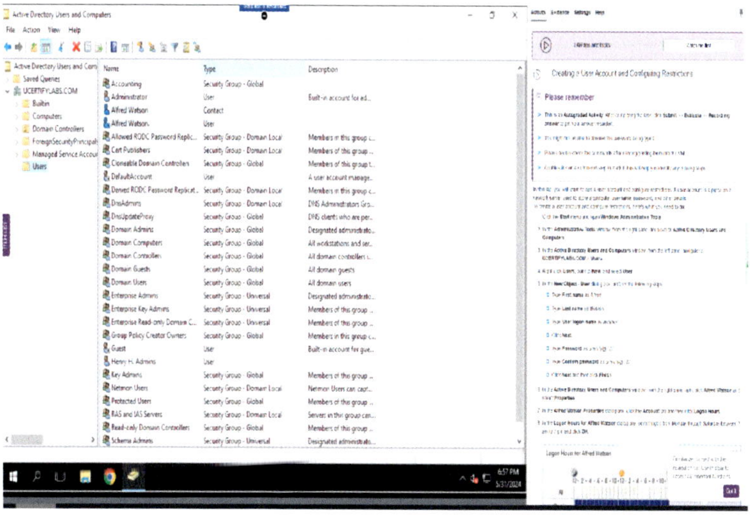

frequently change. Assessors' anticipated knowledge, skills, and duties remain consistent.

SCA Core Knowledge

Before beginning, the assessor must have a clear understanding of specific requirements. The National Initiative for Cybersecurity Education (NICE) Cybersecurity Workforce identifies essential knowledge for SCA work:

1. Understanding network concepts, protocols, and security measures.
2. Risk management processes.
3. Cybersecurity laws, regulations, policies, and ethical considerations.
4. Privacy principles for cybersecurity (best security practices).
5. Cyber risks and weaknesses.
6. Validation and appraisal processes by the organization.
7. Security assessment procedures.
8. Information technology security concepts (firewall, DMZ, encryption).
9. Penetration testing tools and procedures.

When conducting security assessments for information systems and infrastructure, it's important to have a solid grasp of these main topics. Assessors' expertise levels vary based on their job. Some evaluations may need a subject matter expert to consult, represent, or participate in the evaluation to ensure the assessor has accurate knowledge.

SCA Core Skills & Abilities

Assessment scope significantly impacts skills and abilities. If the assessment involves a penetration test, the assessor should have expertise in operating system fingerprinting, port scanning, enumeration, and related techniques. Assessing policies and procedures may need compliance with the NIST risk management framework. Core talents and abilities include the following:

- Demonstrated ability to design effective security systems.
- Demonstrated ability to assess system security controls.

Additional skills and abilities include performing vulnerability scans, protecting confidentiality, integrity, and availability, interacting with customers, preparing test and evaluation reports, and identifying systematic security issues.

- Performing vulnerability scans, protecting confidentiality, integrity, and availability.
- Interacting with customers, preparing test and evaluation reports, and identifying systematic security issues.
- Identifying security concerns through vulnerability analysis.

- Asking clarifying questions.
- Effective communication skills.

The list of skills and expertise on NICCS.CISA.GOV varies according to the extent of the task at hand. A general support system with many physical locations, operating systems, network devices, and procedures requires a comprehensive examination.

Tasks of the SCA1

Assessors must identify assets and their role in the environment before comparing them to a standard. The organization's security policy establishes these criteria for assets. The policy should be based on industry standards and best practices. The assessor's tools include vulnerability, compliance, and wireless scanners. The SCA must understand the nature of the security threat, such as network, system, or physical security, and the assets to defend.

Here are some of the tasks that the CISA mentions:

Conduct a security evaluation to identify weaknesses in security architecture and create a risk management plan.

- Analyze risks.
- Plan and execute security authorization reviews.

- Ensure application software, network, and system security postures are implemented correctly, document variances, and identify corrective measures.
- Evaluate the efficiency of security safeguards per government or industry regulations and laws.

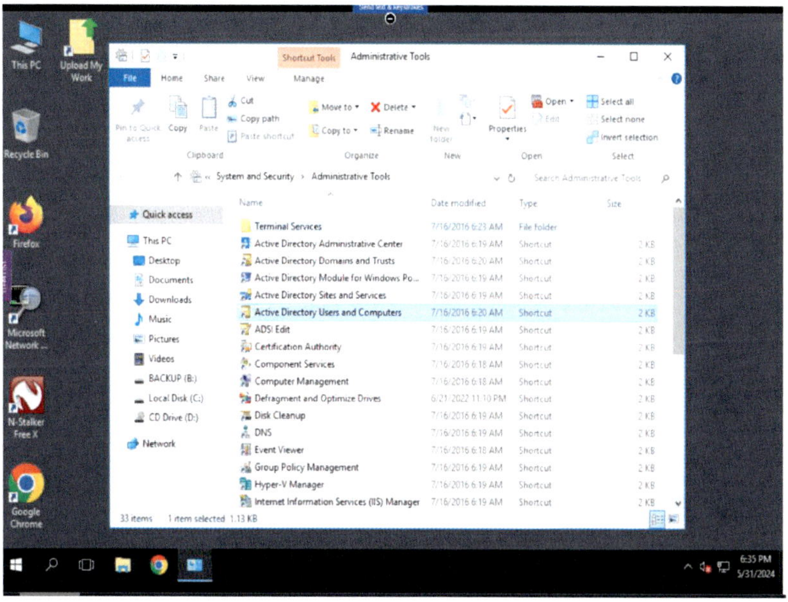

Chapter 3

Security Framework Assessments Process

All leading security frameworks feature an "assess" step, including ISO 27001 (international information security standard), SOX (financial security guidelines), PCI DSS (credit card security rules), and many others. Each has a rigorous evaluation procedure, underscoring that a functional security framework requires an evaluation process to ensure that security measures are in place.

ISO 27001 Evaluation Process

The ISO 27001 evaluation process involves identifying assets, assessing risks, establishing a baseline, and completing the applicability statement. Companies worldwide seek ISO 27001 accreditation, such as Air Maestro in Australia and Nigeria's "You Verify" program. For instance, if you're a Canadian data storage corporation looking to expand into Norway's market, a 27001 certification demonstrates a thorough security process, indicating to potential clients that your firm undergoes frequent assessments.

Assessments and InfoSec for Financial Transactions

Many frameworks serve to secure financial transactions, with SOX and PCI DSS being the most prevalent in the United States.

Sarbanes-Oxley Act (SOX)

Introduced in 2002, the Sarbanes-Oxley Act aims to safeguard investors from misleading financial reporting by firms. SOX covers information security management and risk assessment, involving:

- Identifying and assessing risks.
- Evaluating fraud risks.
- Identifying and assessing changes that may impact internal controls.
- Payment Card Industry Data Security Standard (PCI DSS)

The PCI DSS is a collection of security guidelines established by Visa, MasterCard, Discover, and other credit card companies. It provides recommended practices for organizations using POS systems to gather credit card data. The PCI DSS evaluation procedure involves identifying cardholder data locations, inventorying IT assets, and analyzing potential risks.

NIST 800 Assessment Overview

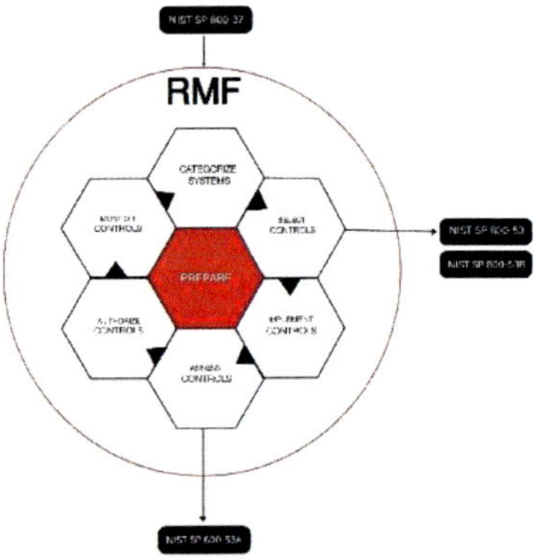

This book focuses on how NIST 800 implements security control assessments, comparable to other security frameworks. The NIST 800 security control evaluation procedure involves:

Assessment Preparation: Determine the type of assessment, necessary resources, and required personnel.

- Develop a Security and Privacy Assessment Plan: Share it with stakeholders.
- Coordinate with Stakeholders: Carry out agreed-upon assessments and document results.
- Analyze Assessment Report Results: Prepare a Security Assessment Report (SAR). The security assessor will review the findings and submit them to the enterprise.
- Post-Assessment Presentation: The assessor will present the findings.

Assessment Regulations

The NIST 800 security control assessment procedure is based on the NIST 800-37 risk management framework and NIST special publication 800-53A. Assessors can learn the full procedure from NIST 800-53, 800-53A, 800-115, and other special publications. All of NIST's special publications are based on government rules OMB A-130 and FISMA 2014.

OMB Circular A-130

The Office of Management and Budget (OMB) Circular A-130 requires that security and privacy control evaluations be based on measures specified by the agency. The evaluation ensures that the controls are properly established, performing as intended, and meeting requirements. Businesses must keep a near-real-time picture of the system's dangers and plan for the most common security issues. Security assessments should be tailored to the value of the information system. For example, a classified weapon system and a public website require different levels of inspection.

Federal Information Security Modernization Act of 2014 (FISMA)

FISMA establishes security standards and recommendations for federal information systems, requiring organizations to perform annual independent reviews for each major system. The assessments evaluate the efficacy of information security policies, processes, and practices. According to FISMA, the Inspector General Act of 1978 mandates the appointment of an Inspector General for certain agencies. Even if an institution is not subject to this rule, the head must conduct an annual independent review of specific systems.

National Security Systems (NSS) store and transmit intelligence and military information to defend the nation. These networks frequently involve more than just money and branding; sometimes human lives, a country's credibility, and other factors are at risk. FISMA mandates that businesses with an NSS conduct assessments by an authorized authority to guarantee information security at an appropriate level. This must be done at least once a year, although the agency's chiefs may require it more frequently.

RMF and Security Controls

Books 1 and 2 of the NIST 800 Cybersecurity series address the RMF method and its accompanying control interpretation. We shall examine the procedure in the context of the assessment.

This book covers all control families from NIST's special publication 800-53, Security and Privacy Controls for Information Systems and Organizations. NIST 800-53A covers security control evaluation methodologies for information systems and organizations. The special publications address all OMB Circular A-130 and FISMA requirements.

The NIST 800 risk management approach does not require a comprehensive examination of all 1000+ controls. During the RMF process, the system only employs controls that are appropriate for its relevance. According to NIST special publications, information systems can be classified based on their impact level, allowing for appropriate security measures. A government finance and security system that pays FBI agents nationwide will have stronger security

Creating a User Account and Configuring Restrictions

Please remember

- This is an **Autograded Activity**. After completing the task, click **Submit** >> **Evaluate** >> **Record my answer** to get your answer recorded.
- You might not be able to observe the password being typed.
- Please double-check the commands after sending/pasting them into the VM.
- Double-click on a completed step to mark it, this will help you identify any missing steps.

In this lab, you will learn to add a user account and configure restrictions. A user account is a place on a network server used to store a computer username, password, and other details.
To create a user account and configure restrictions, here's what you need to do:

1. Click the **Start** menu and open **Windows Administrative Tools**.
2. In the **Administrative Tools** window, from the right pane, double-click **Active Directory Users and Computers**.
3. In the **Active Directory Users and Computers** window, from the left pane, navigate to **UCERTIFYLABS.COM** > **Users**.
4. Right-click **Users**, point to **New**, and select **User**.
5. In the **New Object - User** dialog box, perform the following steps:
   - Type **First name** as *Alfred*
   - Type **Last name** as *Watson*
   - Type **User logon name** as *awatson*
   - Click **Next**.
   - Type **Password** as *ucertify@123*
   - Type **Confirm password** as *ucertify@123*
   - Click **Next** and then click **Finish**.
6. In the **Active Directory Users and Computers** window, from the right pane, right-click **Alfred Watson** and select **Properties**.
7. In the **Alfred Watson Properties** dialog box, click the **Account** tab and then click **Logon Hours**.
8. In the **Logon Hours for Alfred Watson** dialog box, permit logon from Monday through Saturday between 7 am to 7 pm and click **OK**.

protections than a Department of Agriculture workstation with largely public data. The risk management framework guarantees appropriate security safeguards for each system based on the data's relevance. As a result, classified data will be better protected than a web server with publicly accessible data.

RMF Process Steps

The first step in the RMF process is to identify the degrees of security objectives for the system. The security objectives include:

- Ensuring Confidentiality: Restricting access to information.
- Maintaining Integrity: Preventing information manipulation or corruption.
- Ensuring Availability: Ensuring data accessibility for individuals in need.

As a security control assessor, we evaluate if the system maintains confidentiality, integrity, and availability through properly implemented security measures. The organization's security objectives define the system's impact degree and security category. A system can be classified as low, moderate, or high. NIST security rules are determined based on the system's impact level. Below are the security control families that will be discussed:

Control Assessments Overview

The CA control assessment family is designed primarily for assessments in the RMF process. CA-1 includes the policies and processes for control assessments performed by the organization for continuous monitoring, security certification, and authorization. CA-2 is specific to control assessments and addresses the requirement for an organization to undertake frequent assessments.

The CA-2 evaluation consists of:

- Continuous monitoring within the organization.
- Federally mandated annual assessment.
- Tracking changes.
- Experts conducting the evaluation.

The organization guarantees that CA-2 control assessors have the necessary skills and technical competence to create successful assessment plans and assess system-specific, hybrid, common, and program management controls. Required abilities include a general understanding of risk management principles and extensive knowledge and expertise with system hardware, software, and firmware.

At the moderate-impact level, the system will need CA-2(1) independent assessors, including:

- Conducting impartial evaluations with no conflicts of interest.
- Authorizing officials to decide the level of independence.

The NIST 800-53 standard requires an independent assessor to undertake an unbiased assessment of systems. The term "impartial" refers to assessors who have no conflict of interest in the creation, operations, sustainability, or administration of the examined system. Resources that are unbiased to the system can conduct independent reviews within or outside the company.

High-impact systems require security improvements such as CA-2(1) and CA-2(2). CA-2(2) denotes a "special assessment." Special assessments, such as verification and validation, system monitoring, insider threat assessments, and malicious user testing, can improve operational readiness, performance, security, and privacy.

Special assessments were conducted on several military systems we worked on. We conducted evaluations, validation, and verification to ensure proper functioning. We would examine security features by having operational user's process real-world data on the system. Before proceeding, we would conduct a network or configuration scan to ensure all necessary controls were present. The organization can also contact third-party organizations to conduct assessments, CA-2(3), and external organizations.

Reasons for Assessments

Organizations perform assessments for various reasons. New challenges to the company or sector may lead to extensive social engineering and phishing efforts. A thorough network check may be recommended if existing vulnerabilities are discovered within the company. The head of the organization may also request an ad hoc evaluation that has not been planned. Here are some common reasons for assessments:

Continuous Monitoring

Continuous monitoring is an essential component of the NIST risk management framework approach. The organization analyzes its security procedures in the environment through frequent network scans, security checks, and assessments. NIST 800-137: Information Security Continuous Monitoring (ISCM) for Federal Information Systems and Organizations provides comprehensive monitoring recommendations based on FISMA and OMB rules. The organization decides the frequency of monitoring and defines which controls will be monitored and at what frequency. Monitoring is a motivator for conducting assessments, but it is also something that the SCA looks into during independent third-party evaluations. As an SCA, we implement many of the NIST 800-137 recommendations.

Initial Implementation of a System

After implementing security measures, it is necessary to conduct initial system evaluations. Impact assessments are undertaken for proof-of-concept or separated systems from production environments. This prevents serious vulnerabilities from being exploited when the system goes online. The assessment ensures the system functions properly. Creating a system without evaluations is a major error, even for a basic stand-alone system with few risks and exposures.

Major Changes of a System

Switching from Microsoft to Red Hat Linux requires a new examination of the system. If new operating systems are put on the switches and routers, a fresh evaluation may be necessary. Examples include implementing a new firewall and shifting server farms from on-site to cloud infrastructure. This requires reinstalling or double-checking all security measures. Minor upgrades, operating system patches, and new office programs are examples of non-major modifications.

What Will Be Assessed?

Assessments encompass more than just network scans and involve evaluating several components of a company, including but not limited to:

- Physical examinations.
- Policy and procedures.
- Application assessments.
- PCI assessments.
- HIPAA assessments.
- Firewall rule assessments.
- Risk assessments.

- Network evaluations.

Types of Assessments

Misconceptions regarding evaluations include high-level audits by the Federal Inspector General or third-party auditors. Organizations typically perform internal assessments. As previously indicated, evaluations are necessary for continual monitoring and significant adjustments. Cybersecurity personnel often drive many of these modifications in the system.

Self-Assessments

After implementing controls, establishing new systems, or reconfiguring existing ones, a "self-assessment" may be required due to significant changes. This is where the organization evaluates itself. Assessments serve as the organization's eyes and ears. Self-assessments are integrated with many organizational processes, including configuration management, security and privacy impact assessments, incident response, and the system development life cycle.

Assessments are often performed after significant modifications have been made by system engineers or administrators throughout the change control and configuration management process. The cybersecurity team conducts an impact study, which may include a self-assessment, to analyze the risk of implemented modifications.

Post-Incident Assessments

After a system has been compromised, the organization must assess the affected system and any connected systems to determine if the threat has been effectively contained and if the vulnerability remains exploitable.

By following these detailed assessment procedures and understanding the various reasons and methods for conducting assessments, organizations can ensure their security measures are robust, effective, and continuously improving.

Chapter 4

Security Assessment Plan

The NIST Security Control Assessor Process

The preparation phase will make or break the evaluation. As with anything in life, "success favors the prepared." During preparation, the SCA prioritizes the assessment aim and requirements above the control family. The evaluation might focus on the organization's assets, procedures, or documents. Assessment goals include controls, points of contact, locations, missions, and business functions. The preparation process is determined by the organization's requirements.

Here's a brief description of the full SCA procedure:

- **Prepare the Assessment:** Determine the type of evaluation to be performed, what will be required, and who will be needed.
- **Create an Assessment Plan:** Develop a security and privacy assessment plan and share it with stakeholders.

- **Conduct the Assessment:** Coordinate with stakeholders on what has been agreed upon in the plan, then conduct and record the assessment.
- **Analyze the Assessment Report:** Review the findings and share them with stakeholders.

Organizations' Assessment Requirements

The evaluation is based on the needs and requirements of the company. The SCA cannot proceed until they know exactly what is required. If the organization seeks an independent, objective assessor, the following items must be done before the assessment may take place:

- Select the assessor.
- Ensure the assessment aligns with policy.
- Determine scope and objectives.
- Arrange points of contact.
- Inform stakeholders.
- Ensure the SCA has all the necessary information.

The company will rely on appointed cybersecurity personnel to do self-assessments for the system. Before beginning an evaluation, cybersecurity experts should ensure they understand the requirements. Working within the organization will make this easier.

For example, if the firm requires an examination of its disaster recovery and backup systems, the ideal candidate for the task is familiar with them. If a network scan is required as part of the evaluation, the most suitable candidate will be a network assessment subject matter expert.

To choose the appropriate personnel for the evaluation, the company must first understand its surroundings. Technical leaders must have a thorough understanding of the systems and solutions employed. They cannot rely solely on upper-level management to define what will be evaluated and the requirements. They have to rely largely on technical experts.

Ensure the Assessment Fits the Policy

It is the organization's responsibility to guarantee that the evaluation is conducted according to the policy that governs security and privacy measures. This covers the frequency and type of evaluations that may be performed, as well as the requirements of each system.

Sometimes businesses have parts of the company with operating systems that would be disrupted if a scan were performed at specific times of day. Policies may contain provisions requiring specific authorization to enter certain restricted places before the evaluation.

Objectives and Scope

Before conducting the assessment, the company must ensure that the systems are ready and at the appropriate stage of the risk management framework. This refers to the security controls. Before the evaluation, classification, selection, and implementation (as per NIST 800-37) should be completed.

Knowing the system's present state, security posture, and environment will assist the business in determining the aim and purpose of the required evaluation. Is it a normal continuous monitoring evaluation that simply needs in-house cybersecurity personnel? Is the examination limited to a documentation review? Is it required to conduct an inspector general evaluation across all agency systems?

The organization must determine what exactly needs to be examined. This includes the scope and range of IP addresses that will be scanned, the individuals (if any) that will be questioned, and the documents that will be evaluated for the system. As an SCA, having this information is essential before beginning.

Arrange Points of Contact (POC)

The organization will need to designate points of contact for each set of assets and controls that will be covered. These points of contact will be the persons who work on the assets daily or have access to the resources required by the SCA to perform their duties. This includes internal groups such as common control providers. For example, a common control provider might be the physical security team that is outsourced to regulate access to the site. If this organization is required for the SCA's evaluation, it must be listed as a POC.

Inform All Stakeholders

The company will need to contact all stakeholders impacted by the evaluation. System administrators and engineers may need to serve as points of contact for the SCA or ISSO conducting the evaluation. System operators may also need to be notified if the evaluation would affect their ability to carry out everyday missions and activities. Management will need to assist in spreading the word about the evaluation. Informing all necessary parties will ensure that no one is surprised and that relevant subject matter experts are prepared for any questions or tests that the ISSO or SCA may ask during the evaluation.

Ensure the SCA Has Everything They Need

The business can reduce miscommunications by permitting direct interaction with the subject matter expert managers or representatives who will be required during the evaluation. These POCs will provide the relevant documentation, meet with the assessor to identify any complicated issues that may surface during the assessment, and provide further insight into the systems being examined if necessary.

Assessor / Assessor Team Preparation

Before the assessor can begin the security assessment plan, they will need the following:

- Understand the concept of operations.
- Understand the organizational structure.
- Arrange points of contact.
- Gather artifacts.

Understand the Concept of Operations for the System

The assessor needs to have a good grasp of the system's goals, activities, and business procedures. This will provide context for why some things are the way they are. The company may require a legacy system to fulfill an older purpose that is being superseded. A unique configuration may be employed due to a specific business procedure required by the company. Knowing why the system exists, as well as seeing high-level network diagrams and system security plan data, will provide the SCA with an understanding of its structure.

Organizational Structure

The SCA should understand the organization's entities responsible for developing and implementing the controls and supporting systems.

Points of Contact

Meeting with the necessary organizational officials and points of contact allows the SCA to understand what the system requires, the sort of assessment, the scope, and the level of complexity.

Artifacts

The SCA will need certain artifacts before they even start. These include (but are not limited to):

- Policies.
- Procedures.
- System security plan.
- Designs & diagrams.
- Previous assessment results.
- Security configurations of relevant systems.
- Configuration of security tools such as firewall and router settings.
- After-action reports.
- Minutes from configuration management meetings.
- Authorization packages.
- Ongoing maintenance of controls.

Self-Assessments Preparation

Self-assessments will be the most common type of system evaluation. Internal cybersecurity professionals do self-assessments as ISSOs, GRC policy subject matter experts, or SCAs. To focus on the assessment job, we shall refer to the cybersecurity expert doing this work as an SCA or an assessor.

All of the organization's standards listed above still apply. However, the internal ISSO and SCA will find it much simpler because they are so directly related to the organization's goals and business activities.

During a self-assessment, the SCA will know where to find all of the required artifacts, who to contact for most systems, and where all of the skeletons are located.

If you are conducting a self-assessment, you will still require consent from management inside the business, and stakeholders must be aware of what is going on. Aside from the amount of access provided during assessment preparation, self-assessments are often more casual. Independent third-party evaluations often have defined templates that must be followed, a certain number of meetings with important stakeholders, and expected deliverables at the end of the security control assessment. The meetings are quite formal.

In contrast, a self-assessment will include only the meetings and deliverables required to complete the assignment. If a self-assessment is necessary for configuration changes, the meetings may just cover normal configuration and change controls. I have conducted self-assessments and evaluated the security control statements in the System Security Plan (SSP). In many circumstances, my sole coordination included one-on-one meetings with a system administrator to verify that the data supplied in the SSP was still correct.

Self-Assessment for Major Changes

Configuration management and change controls may necessitate a self-assessment to ensure that security safeguards are still in place and effective. The system administrators, developers, or engineers responsible for making the modifications will be required. The required deliverables are as follows:

- Scans before and after adjustments.
- Network vulnerability assessment.
- Plan of action and milestone papers (if applicable).
- Meeting minutes.
- Impact analysis.

The requirements will differ from organization to organization, thus the assessor must conduct their study.

Self-Assessment for Security Incidents

Depending on the security incident, the organization may require a self-assessment. The self-assessment will determine whether the threat can still exploit the vulnerability. The incident response team will be necessary points of contact. The deliverables may include the following:

- Scans that allow a before and after status (if necessary).
- Incident response report.
- Network vulnerability scan.
- After-action report.
- Forensics data of the breach (if any).
- Root cause analysis report.
- Analysis of audit logs (if any).
- Self-Assessment for Continuous Monitoring

One of the most casual self-assessments is used for continuous monitoring. Assessments are an essential aspect of continuous monitoring. Some self-assessments for continuous monitoring focus on controls rather than assets.

The required points of contact will differ for this type of assessment. You may need to communicate with network administrators, server administrators, or privacy managers. It depends on the set of controls being analyzed. Sometimes, the ISSO performing the SCA must notify management that they are examining controls. They analyze implementation statements in governance, risk, and compliance databases or content management systems (e.g., eMASS, Archer, Xacta). Managers or senior executives responsible for information system security must typically be aware of what is going on.

Some of the artifacts needed for continuous monitoring include:

- Updated system security plan.

- Conducting a privacy effect study (if necessary).
- Conducting network vulnerability scans and evaluating documentation.

There are too many objects and jobs to list here. It's better to focus on the forms of continuous monitoring that will determine which artifacts are required.

Types of Continuous Monitoring Tasks

Self-Assessment for Continuous Monitoring

One of the most casual self-assessments is used for continual monitoring. Assessments are an essential aspect of continual monitoring. Some self-assessments for continuous monitoring focus on controls rather than assets.

The required points of contact will differ for this one. You may need to communicate with network administrators, server administrators, or privacy managers.

It just relies on the set of controls being analyzed. Sometimes, the ISSO performing the SCA must notify management that they are examining controls. They analyze implementation statements in governance, risk, and compliance databases or content management systems (e.g., eMASS, Archer, Xacta). Managers or senior executives responsible for information system security must typically be aware of what is going on. Some of the artifacts

- Update system security plan
- Conduct a privacy effect study (if necessary).
- Conduct network vulnerability scans and evaluate documentation.

There are too many objects and jobs to list here. And it's better to focus on the forms of continuous monitoring that will determine which artifacts are required.

Many activities and strategies can be utilized to continuously evaluate an organization's self-assessment. Here are a few:

Social Engineering:

- Internal Phishing Campaign.
- Physical Security Checks and Challenges.
- Cyber Threat Intelligence.
- Penetration Testing.
- Application Security Scans:
- Web application assessment.
- Source code evaluation.
- Cloud Assessments.
- Verification and Validation.
- Assess Policies, Procedures, and Processes.

Chapter 5

Security Assessment Plan

Assessors or those responsible for SCA (Security Control Assessor) duties create the security assessment strategy, also known as a security and privacy assessment plan (SAP) or simply an assessment plan. According to the NIST Computer Security Resource Center, an assessment plan describes the objectives for the control evaluation and provides a detailed roadmap for performing the assessment. This strategy is tailored to the organization's unique needs and requirements. While the SCA can follow a basic template for an SAP, there is no one-size-fits-all solution. The security and privacy strategy will address the assessment's "who, what, when, where, why, and how." The SCA documents the strategy, which is then reviewed and approved for the organization's independent third-party evaluations.

Overview of a Security Assessment Plan (SAP)

The security assessment strategy often includes privacy restrictions. If necessary, the organization might develop a separate evaluation strategy for privacy. Systems that process, store, or transmit personally identifiable information (PII) will require a separate privacy procedure consisting of a Privacy Threshold Analysis (PTA) and a Privacy Impact Analysis (PIA), which will be documented and updated regularly as defined by the organization. The privacy assessment procedure is often a self-evaluation conducted by the company. However, the SAP does not exclude any privacy protections linked to the system under review. Privacy concerns will be documented in the security assessment report if they arise.

The SAP outlines the objectives for the security and privacy control evaluations and provides a roadmap. Depending on the organization's demands, assessment plans might be integrated or distinct. For example, there may be a separate assessment plan for network scans, another for physical security assessments, and one for reviewing policies and procedures. During the preparation stage, the organization and SCA will decide what goes into the plan. The assessor and organization should consider the following stages when formulating strategies to evaluate security and privacy controls:

Determine which assets and sites will be assessed within the system's authorized boundary.

Identify which security and privacy controls, as well as control enhancements, should be included in the assessments based on the security plan and privacy plan (or equivalent document for non-system-based common controls) and the assessment purpose and scope.

- Choose suitable assessment techniques depending on security or privacy protections, including upgrades.
- Select assessment techniques, including methods, objects, and depth/coverage attribute values.
- Create new assessment methods to satisfy security needs or measures not specified by NIST Special Publication 800-53.
- Optimize assessment techniques to eliminate redundancy and offer cost-effectiveness.

Security and Privacy Assessment Plan Format

Some security assessment plans begin with a cover page. The cover page can include the organization's name, the system's name, the document's version, and the date. The cover for the assessment plan is optional and depends on the organization's preferences and format. If the SCA is from an upper-level agency, it will follow a specific structure for each lower-level entity to which it is delivered. Self-assessments seldom require a cover page, and the structure will be as straightforward as possible.

Formats I've seen include a document created by a governance, risk, and compliance database/content management system, such as RSA Archer or eMASS, which includes the system's name, the date the assessment will take place, and the controls scheduled to be examined. And that's it.

Security and Privacy Assessment Plan Purpose

Given the numerous causes, occasions, and events that might trigger an assessment, it is critical to specify the goal of the assessment within the SAP. This could be part of regular internal continuous monitoring, a yearly IG audit, or a phase in the change control process. Whatever it is, it should be included in the SAP notes so that others working on the document may follow up, and upper-level management can distinguish this evaluation from the rest. Remember, organizations perform various assessments and ratings on their business and mission systems, so the more explanation you provide, the better. The purpose also provides context for the assessment.

Sample Purpose Statement (Self-Assessment)

The goal of this security control evaluation is the yearly continuous monitoring set by [Organization A] for [System A]. This follow-up assessment focuses on the remaining NIST 800-53 security controls and evaluates all associated risk response paperwork, including acceptance, plan of action, and milestone documents.

Sample Purpose Statement (Third-Party Assessment)

The Department of Homeland Security (DHS) will conduct an independent audit of [Organization A]'s business and mission systems. External network connections, IP addresses, and hostnames shall be detailed in Appendix AA of the security assessment plan.

Security and Privacy Assessment Regulations & Background

This section of the SAP will include a summary of the key federal, state, and industry rules that govern and guide the assessment process. Federal security assessment plans may mention FISMA and OMB. Any applicable national, county, or state information system rules, as well as any required industry regulations, would be referenced. Examples include HIPAA for healthcare, PCI DSS for firms that accept credit cards, and ISO 27001 for international systems.

Sample Security and Privacy Assessment Regulations

The organizational security control assessment is governed by the Federal Information Security Modernization Act of 2014 (FISMA) and the Office of Management and Budget (OMB)

Circular A-130 laws, which were enacted to secure government information systems.

Security and Privacy Assessment Process and Methodology

This section describes the assessment technique for verifying and validating that managerial, operational, and technical controls are properly applied. The assessor may provide a summary of network scanning technologies or explain the use of penetration testing. To evaluate management and operational controls, the assessor must indicate that interviews and document reviews will be part of the process.

Security and Privacy Assessment Scope

The scope of the SAP is crucial. This part can contain the name and description, but it is more important to describe the assessment's boundaries. It specifies what gets scanned and where, assuming there are many physical locations. This section may include a list of all the information systems or devices that were evaluated. The scope can include a separate appendix that identifies all the systems to be evaluated, mentions the range of IP addresses, or simply refers to the systems if all the individual systems are clearly defined and identified by the business. Before the assessor can proceed, the organization must agree on the scope.

Assessment Assumptions and Limitations

Assumptions are expectations accepted without question during the security assessment process. These might include the assessment scope, team knowledge and abilities, resource availability, and documentation accuracy. It is critical to specify these assumptions in the strategy so stakeholders understand the foundation for the evaluation and can judge the validity of the outcomes.

Limitations, on the other hand, are factors that may influence the accuracy or completeness of the evaluation. These may include technological limits, such as the inability to access particular systems or data, as well as organizational limitations, such as prohibitions on the use of specific evaluation tools or methodologies. It is necessary to recognize these restrictions, as they can impact the assessment results and should be considered when evaluating the findings.

It is vital to emphasize that the assumptions and limitations section is not designed to explain any flaws in the evaluation method. Instead, it provides transparency and context for the evaluation, ensuring stakeholders understand the extent and bounds of the study.

In conclusion, the assumptions and limitations section of a security assessment plan is an important element of the document because it offers clarity about the process and specifies any constraints or boundaries that may impact the accuracy or completeness of the assessment. It is critical to clearly explain assumptions and limitations in the plan so stakeholders can assess the validity of the outcomes.

Chapter 6

Conduct the Assessment

With an agreed-upon security assessment strategy in place that defines the who, what, when, where, and how of the assessments, it's time to get to work. This can be done on-site, remotely, or as a combination of both. For example, the network scanning and documentation review sections of the examination (if required) can be completed remotely, while the physical security check and wireless evaluation will be conducted on location. The method depends on the specifics outlined in the security assessment strategy. Conducting as much of the assessment remotely as possible is often cheaper, as on-site inspections require significant time, money, and resources.

What Professional Assessors Do

Managing expectations and staying within the assessment's scope distinguishes professional assessors from amateurs. Allow yourself more time than necessary to complete the evaluation. If you anticipate spending two days on-site, request three or more days and attempt to do as much work as possible before you arrive. Do not exceed your required number of days. If you ask for two days and then need three and a half, the client is likely to complain. Every minute they guide you costs them money; it is preferable to request more time than you require and complete the task in less time. Always stay within the parameters of the evaluation. One of the worst things you can do is examine systems without following the security assessment plan. Discovering significant vulnerabilities on systems you weren't supposed to scan can create serious issues and undermine your client's trust.

Conducting Self-Assessments

Self-assessments conducted by in-house cybersecurity specialists are significantly more casual. Communication with stakeholders is simpler, and there is less pressure, allowing the assessor to design an assessment timetable that aligns with expectations more easily. Self-assessments are typically prompted by routine internal operations, such as normal continuous monitoring or large system changes, making them calmer and more internally focused. They will most likely only be viewed internally. This is significantly different from independent third-party assessments, which are pushed from higher levels and receive less scrutiny since the outcomes are organized. However, the same methodologies and procedures may be utilized for both self-assessment and third-party evaluations.

Assessment Methods

The NIST 800-115, Technical Guide to Information Security Testing and Assessment, is a useful reference for the SCA. According to 800-115, an information security assessment evaluates how well an entity (e.g., host, system, network, method, person) achieves security goals. In other words, a security control evaluation examines how successfully a system or organization's security elements are implemented. To do this, the assessor examines items within the organization and system. The NIST 800-53A refers to the items you will study as "objects." Security rules, processes, host system configurations, and interviewees are all considered "objects." There are three approaches for assessing security control implementation:

- Conduct examinations
- Conduct testing
- Conduct interviews

The SCA analyzes, tests, and interviews objects (documents, systems, and organizational resources and assets) to evaluate whether security measures have been properly applied. The SCA examines what the organization has in place and compares it to the assessment goals. Assessment goals are what the assessor should look for in terms of compliance or expected outcomes. Expected results might be found in NIST 800-53A, the organization's standards, best security practices, or a combination of these.

Assessment Method: Examine

The "examine" assessment process entails observing, verifying, reviewing, inspecting, analyzing, or assessing items. In this context, "objects" include documentation, host system settings, information systems, processes, people, networks, and any other resources that enable security control. The examination approach is used in almost every security control evaluation, whether looking at policies or assessing test results to establish control efficacy.

Assessment Method: Interviews

Interviews are interactions with persons who are familiar with or have worked with the security control system in some capacity. An interview is a more effective way to analyze specific operational security measures than a test or examination. For example, an interview would be the best way to determine how well subject matter experts understand their procedures. Some queries might provide a more in-depth understanding of how the company and its systems are safeguarded. When users, subject matter experts, and system managers make specific remarks, it is possible to quickly cover a lot of territory for assessment. Their assertions are known as "attestations" and can be utilized to pass or fail a security control implementation.

Assessment Method: Testing

Testing is ideal for technical system controls. This is where the assessor interacts with the information system to observe behavior. Normally, the SCA will have the subject matter expert (system administrator, firewall administrator, network engineer, etc.) log into the system and enter the required commands. This is the ideal circumstance since the subject matter expert is more familiar with their systems and is best suited to technical system controls. Testing includes logging in, running system restorations and backups, accessing configurations with various accounts, viewing event logs, showing firewall rules, proving system capabilities, and much more. Any direct engagement with the system to determine the outcome is part of testing.

Expected Results

Regardless of the technique used to assess the efficacy of the control, the assessor compares the examination, testing, or interview results to what the organization is supposed to be doing. What they are meant to perform is documented in the organization's policy. If there is no policy, strategy, or document outlining the constraints and rules, they are codified in federal, industry, state, or local regulations, directives, or laws. We will go into deeper detail regarding the assessments for each control household in subsequent chapters.

Chapter 7

Security Assessment Report (SAR)

The Security Assessment Report (SAR) includes all the conclusions from the security control evaluation. It provides an executive summary of what was done, an overview of the overall risk level, identifies each vulnerability discovered, and offers recommendations on how to address each finding. The SAR can be presented in various formats, such as a spreadsheet, PDF, secure web page, or even a PowerPoint presentation, depending on the organization's preferences. A more extensive examination will provide a complete package that may contain hundreds of pages as well as network scan data. Regardless of its length, the SAR must include a summary for upper-level management, an overall risk assessment, and an analysis of each vulnerability.

Importance of the SAR

The importance of timely SAR delivery cannot be overstated. Once, I was late providing an assessment report to two of my clients.

Client A was not in a hurry as they needed more time to address some known weaknesses. However, Client B urgently needed the report for internal quarterly reviews tied to corporate bonuses. Client B's dissatisfaction led them to contact both the CEO and my boss. The lesson here is clear: never be late with your reports. Allow yourself plenty of time to conduct the assessment and compile the report.

Assessments are vital for organizations, functioning as their eyes and ears. Without them, they are navigating in the dark. For Client B, the evaluations were crucial for determining risk management funding. This underscores the SAR's role in supporting organizational decision-making and resource allocation.

Layout of a Typical SAR

I. Executive Summary

- Overview of the risk assessment process
- Key results and recommendations for risk mitigation
- Conclusion and future measures

II. Introduction

- Purpose and scope of the risk assessment
- Methodology and approach
- Assumptions and limitations

III. **The Risk Assessment Process**

- Preparing for the evaluation: defining its scope and objectives, identifying assets and stakeholders, and determining risk tolerance criteria
- Conducting the evaluation: identifying and assessing risks and vulnerabilities, analyzing possible impact and likelihood of occurrence, determining the amount of risk, and prioritizing actions
- Maintaining the assessment: continuously monitoring and updating the risk assessment, incorporating feedback and lessons learned, and communicating risks with stakeholders

IV. **Results & Findings**

- Summary of detected risks and vulnerabilities
- Analysis of possible effects and likelihood
- Prioritization of risk mitigation suggestions
- Assessment of residual risk following mitigation

V. **Conclusions and Recommendations**

- Risk level assessment
- Key findings and recommendations
- Action plan for risk management and reduction

VI. **Appendices**

- Detailed methodology and analysis

- Supporting documentation (risk matrices, threat models)
- Glossary of essential terminology and acronyms

Chapter 8

Post-Assessment

Post-assessments are not documented in the NIST 800 series, but in my experience, they are conducted at practically every organization. After the assessment is completed and a security assessment report is created, a process of review, SAR adjustments, and distribution is expected to occur.

For self-assessments performed as part of continuous monitoring, the post-assessment may be as easy as examining the results, obtaining peer review, and forwarding them to the appropriate management for additional examination. 'This is dependent on the level of self-assessment and internal processes. However, doing evaluations for third-party entities is different.

The post-assessment process for third-party evaluations, in which you act as an impartial assessor for an organization, is more complicated.

- Always let someone else on the team evaluate it. This is called a peer review, and it is extremely essential.
- Deliver the report to the team for feedback.
- Give them a few days to review the results and identify any
- Inconsistencies.

- Review the organization's reactions and comments.
- Present results to stakeholders.

The assessment report is sent to the organization, and the organization that administers and owns the system being reviewed has the opportunity to reply. 'The opportunity to respond must be included in the report. Everyone must understand that the report is not final until the client's security team has.

Creating a User Account and Configuring Restrictions

Please remember

- This is an **Autograded Activity**. After completing the task, click **Submit** >> **Evaluate** >> **Record my answer** to get your answer recorded.
- You might not be able to observe the password being typed.
- Please double-check the commands after sending/pasting them into the VM
- Double-click on a completed step to mark it, this will help you identify any missing steps.

In this lab, you will learn to add a user account and configure restrictions. A user account is a place on a network server used to store a computer username, password, and other details.
To create a user account and configure restrictions, here's what you need to do:

1. Click the **Start** menu and open **Windows Administrative Tools**
2. In the **Administrative Tools** window, from the right pane, double-click **Active Directory Users and Computers**
3. In the **Active Directory Users and Computers** window, from the left pane, navigate to **UCERTIFYLABS.COM** > **Users**.
4. Right-click **Users**, point to **New**, and select **User**.
5. In the **New Object - User** dialog box, perform the following steps:
    - Type **First name** as *Alfred*
    - Type **Last name** as *Watson*
    - Type **User logon name** as *awatson*
    - Click **Next**.
    - Type **Password** as *ucertify@123*
    - Type **Confirm password** as *ucertify@123*
    - Click **Next** and then click **Finish**.
6. In the **Active Directory Users and Computers** window, from the right pane, right-click **Alfred Watson** and select **Properties**.
7. In the **Alfred Watson Properties** dialog box, click the **Account** tab and then click **Logon Hours**.
8. In the **Logon Hours for Alfred Watson** dialog box, permit logon from Monday through Saturday between 7 am to 7 pm and click **OK**.

MAX
RISK
LEVEL
MIN

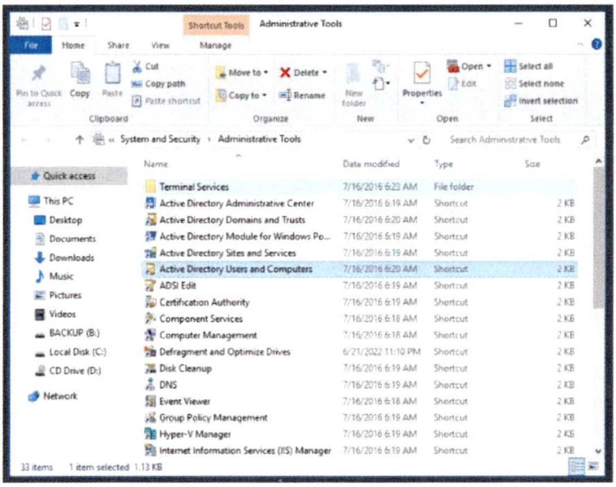
Shortcut Tools
Administrative Tools
File
Home
Share
View
Manage
Pin to Quick access
Copy
Paste
Cut
Copy path
Paste shortcut
Clipboard
Move to
Copy to
Delete
Rename
Organize
New folder
New
Properties
Open
Edit
Open
Select all
Select none
Invert selection
Select
System and Security › Administrative Tools
Search Administrative Tools
Name
Date modified
Type
Size
Quick access
This PC
Desktop
Documents
Downloads
Music
Pictures
Videos
BACKUP (B:)
Local Disk (C:)
CD Drive (D:)
Network
Terminal Services
Active Directory Administrative Center
Active Directory Domains and Trusts
Active Directory Module for Windows Po...
Active Directory Sites and Services
Active Directory Users and Computers
ADSI Edit
Certification Authority
Component Services
Computer Management
Defragment and Optimize Drives
Disk Cleanup
DNS
Event Viewer
Group Policy Management
Hyper-V Manager
Internet Information Services (IIS) Manager
33 items
1 item selected 1.13 KB

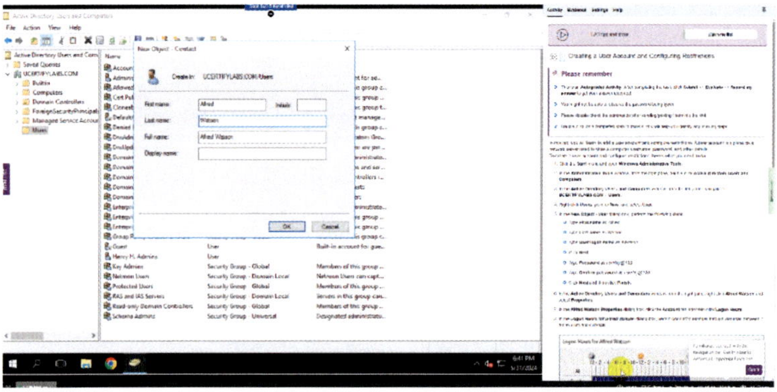

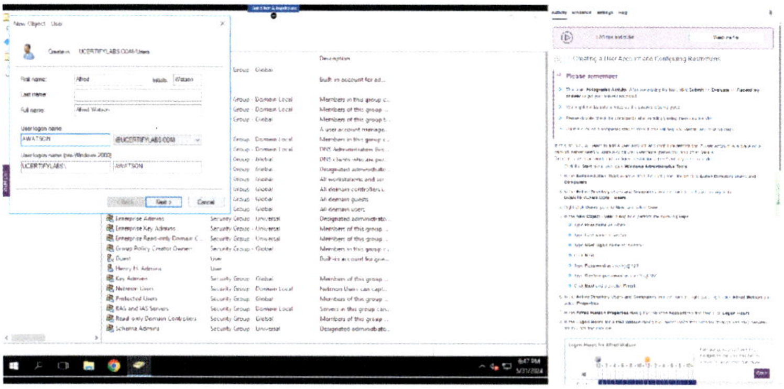

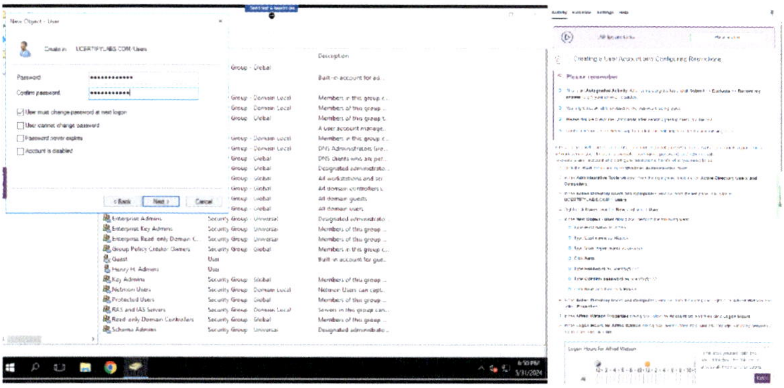

Responding to False Positive

A false positive occurs when a scanner discovers a fault that does not exist. The vulnerability scans discover flaws using a predetermined technique that extracts data from system files or compares the version of a program found on the system to the most recent secure version that should be on the system (for example).

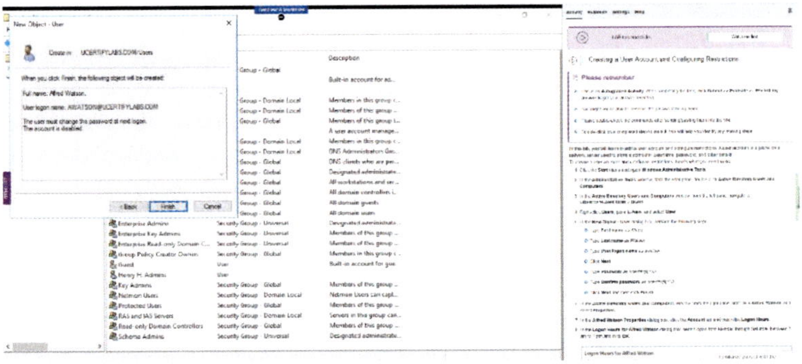

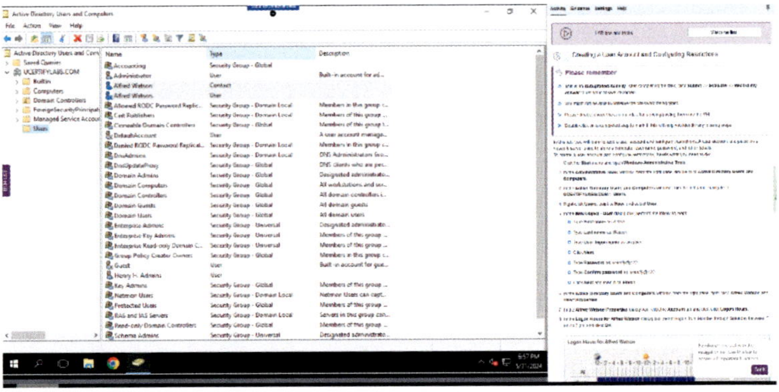

False positives from vulnerability scanners might include flaws that cannot be exploited, misidentified software or operating systems, and system settings that the scanner misinterprets as security issues.

As an assessor, you may not catch all of the false positives. Your primary goal is to discover flaws across numerous systems; you have a limited amount of time to do this.

As an assessor, you evaluate the findings based on whether the scan was completed successfully on the tight systems. You may not have time to delve deeply into each discove ry. The assessor must provide the cybersecurity team and system owners time to analyze the results and explain Gf (if necessary). The cybersecurity team and system owners are the most familiar with their systems, thus they will be aware that a certain patch was updated and included in the assessor's assessment.

Known Documented Issues

If the company has been doing self-assessments as part of its continuous monitoring, it will generate a risk response for issues that could not be resolved within the mandated vulnerability remediation time frame. A risk response acknowledges the danger and explains how the company will address it. There are a variety of risk responses:

- Plan of Action and Milestones (POAM) A POAM is a document that defers the fixed actions of weakness by developing a plan that is tracked throughout time.

- Risk acknowledgment documents, often known as "risk acceptance," declare that the company is aware of the risk and provide a rationale for its existence. They have reduced the danger by securing the system with alternative methods.

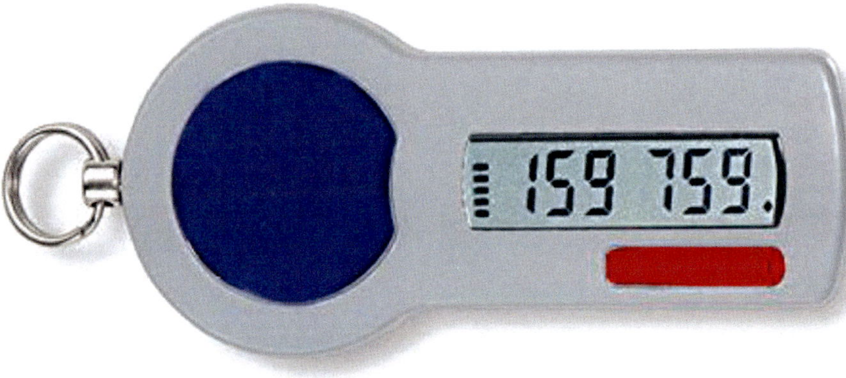
159 759.

Post-Assessment Meeting

Meetings are not always required; internal self-assessments do not often include a meeting. If there is a meeting, the assessor should present the major findings and include any reactions from the organization.

After everything is finished and the organization has replied, the SCA may meet with all of the key stakeholders. The major points of contact will be the cybersecurity team, IT management (if accessible), and a C-level executive. The C-level executive should be the Chief Information Officer, Chief Information Security Officer, Authorizing officer, or any other person who signs the authorization package. Because this is who will receive the report.

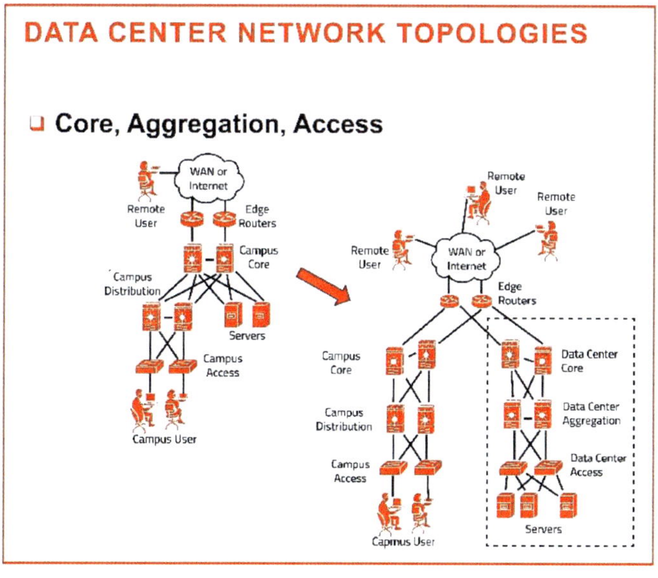

AC - Access Control Family Assessment

When looking at the AC, Access Control, family from the standpoint of a security control assessor, we must determine if the company controls logical access to its assets. Our initial step is to determine whether assets provide a business-critical function. We next want to verify if they safeguard network access, wireless access, remote access, and mobile phone access to those vital tasks.

This control family will take longer than usual since numerous controls need account administration. For example, if badging is associated with an account, physical and personnel security (PE and PS controls) will be determined by account management, whereas audit logs (AU controls) demand that particular occurrences identify the account. Other controls that rely on account management are identity, authentication, and privacy.

Identification, authentication, and privacy controls are only a few examples of account management-based controls.

As Assessors, we would want to examine the following controls in the AC family:

- AC-1 Policy and Procedures
- AC-2: Account Management
- AC-17 and Remote Access (if appropriate)
- AC-18 and wireless access (if available).

I am not advising you to ignore any other variables in the AC family. Lam simply stated that, at the very least, we want to evaluate these four keystone controls, which provide insight into the organization's overall access control policies.

Identify the Essential Functions for Access Control

The focus is on the access control procedure. The emphasis is not on a golden server in a restricted area with sensitive information that performs all necessary operations. Focusing on the access management process will discover concerns on all systems that perform vital activities. However, knowing where the most important mission necessary functions are will allow the assessor to determine the amount of risk to the company.

As a SCA, we are not obsessed with just one server or even a collection of controls. We may utilize a sampling of systems across the company, and keywords; we are searching for the organization's access process. We simply do this to monitor the behavior that the organization has set up on the systems.

We evaluate a set of key DNS servers or a crucial database. Realistically, if the business lacks access restrictions, an attacker may get access to mission-critical functions from an end user's machine via wireless, remote access, or other means.

Examination of Account Management Controls

The account management process should be documented, explicitly specified in a policy, and described in a procedure. The policy must be updated and signed by higher management. Just be sure the insurance has not expired. If it is three years old or signed by leadership who is no longer in place, include this information in the risk assessment report. The procedures should clarify how access management is being carried out. Procedures may make use of obsolete instruments and procedures.

At a minimum, the policy should address the following:

- Role-based access (AC-3)
- Account creation and termination (AC-2).
- Monitoring accounts
- Detecting unsuccessful logon attempts
- Notifying users about system use

Account management policies and procedures may involve many papers. Documents may contain overall security policies, access control policies, and personnel termination or transfer policies (for human resources).

Other Evidence of Account Management

The SCA may also see a list of active and disabled system accounts, as well as a list of groups and their members. Active Directory and other account management technologies enable system administrators to export accounts.

Interview for Account Management Controls

You should familiarize yourself with the account management process before speaking with the point of contact. This might be a cybersecurity specialist, a system administrator, or someone knowledgeable about the organization's account management. In my experience, the best persons to talk to are those who create, manage,

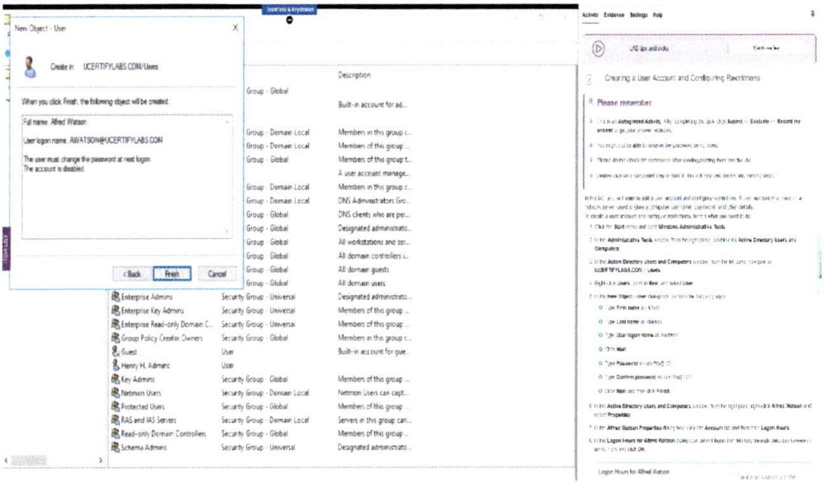

and delete accounts. Their knowledge is so extensive that after only a few questions, I have nothing further to ask. Here are some questions to ask regarding access management:

- What account management applications do you use (e.g. Active Directory or LDAP)?
- What account types do you have (administrator, guest)?
- How and when do accounts get removed?
- How do new users create an account?
- Does the security staff have?

All of these questions assume that the firm has not only an account management system but also a procedure in place. If this is implemented well and the SCA speaks with the point of contact who understands how it works, they will be simple "softball" questions. If the SCA speaks with someone unfamiliar with the account management procedure, the organization does not have one. They will begin to sweat and stutter, and it will feel as if you are questioning them.

Testing of Account Management Controls

The SCA can determine if the company employs access management. One method is to log in to a system that allows you to show the use of your account.

Several approaches may be used:

- Log in as a guest.

- Access administrative tools as a guest.
- Log in as a normal or privileged user.
- Log in with no credentials.
- Log on with random credentials.

The expected test results are determined by the organization's security policy.

Red Flags of Access Management

If you see a large number of accounts from people who have left the company, this is a major issue. It indicates that the organization's approach may not include account deactivation. If these accounts are not terminated, they become forgotten and vulnerable to assault.

Green Flags of Access Management

It is quite beneficial for a business to have a regular accounting review procedure. Every account has a tight lifespan, and any loose ends are collected during the monthly inspection.

Examination of Wireless Controls

This only applies if the system being reviewed is accessible via wireless networks or has Wi-Fi or Bluetooth enabled. There will need to be wireless limits.

Wireless restrictions will be established in a policy and process. Wireless documentation should provide configuration and connectivity needs. An example is a policy that states the necessary use of WPA-Enterprise and limits on guest accounts, as well as the usage of Bluetooth. Other papers should guide the implementation.

Bluetooth is sometimes ignored and disregarded. Users may bring their own Bluetooth devices and link them to the organization's systems, therefore prohibitions in writing are necessary.

Frequently, organizations and individuals are ignorant of typical Bluetooth breaches and vulnerabilities, including:

• Blue jacking involves blasting a message to all Bluetooth-enabled systems.

• Blue bugging involves creating a backdoor on a phone or laptop over Bluetooth.

• Blue Borne, an assault device (Armis) infects the system via Bluetooth.

• Bluc Snarf connects to a system to steal data.

The organization should have a policy that restricts what may be linked to its infrastructure. An organization that is completely aware of Bluetooth hacks will have some of these limitations documented:

• Turn off Bluetooth when not in use.

• Do not connect with untrusted devices.

• Do not save sensitive information on a Bluetooth device.

• Use a strong password or PIN.

• Update to the latest Bluetooth version.

• Enable Bluetooth security features.

The Wi-Fi policy should address typical vulnerabilities, including default router setups, rogue access points, SSID protection, and WEP use.

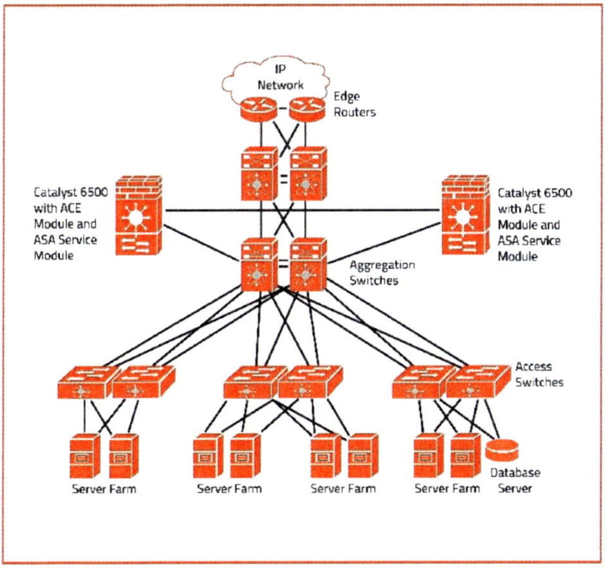

Testing of Wireless Controls

Many evaluations contain components that may be performed at a single location or even remotely to determine much of what is going on at other sites. Remote capabilities include network scanning and document assessments. However, much like physical examinations, wireless testing must be performed at each site. The wireless range and overlap with other wireless broadcasts were all considered in the assessment.

The wireless evaluation consists of three parts:

Step 1: Talk with the wireless owners.

Step 2: Conduct the wireless. Internal & External

Step 3: Collect data.

Step 1: Talk to the Wireless owners

Before you stroll around the organization's site using wireless scanning equipment, you must-obtain consent from top management and the place you will be visiting. The SCA and the target organization must agree on the scope and needs of the assessment for the specific location. An SCA should never undertake a wireless evaluation without management's written approval and the individual site's acceptance. There may be particular setups that you should be aware of. For example, at one location I visited, the organization had a partner organization with wireless and Bluetooth-enabled devices on the facility's north side, one of which required prohibition of data from the other wireless network. The site manager informed me of these data, which were not included in the overall security assessment plan.

Step 2: Conduct the wireless internally and externally

You can scan using a variety of devices and applications. These include the following, which will vary with time:

- Storm pad is a CEH Raspberry Pi device that scans all Wi-Fi and Bluetooth devices on the network.
- Pwn Pad is a wireless and Bluetooth scanning device.
- Blue Hydra is a Bluetooth scanning application.
- Wi-hawk is a tool for auditing IP addresses on wireless routers.
- Bitdefender Home Scanner scans Wi-Fi networks.
- Fing App is a mobile app for network scanning.

There are several wireless scanning devices and applications that may be utilized on the organization's websites. Let the customer know which tools will be utilized.

Whatever scanning solution you use, make sure it can collect data on all SSIDs, Wi-Fi access points, and any Bluetooth vulnerabilities throughout the facility.

You will need to build a wall around the inside of the facility, protected sections, and the outside boundary. You will also need a point of contact to accompany you to restricted regions. Do not do Bluetooth scans outside; Bluetooth has a 10-meter range, therefore if you conduct a Bluetooth scan outside the facility, you will pick up items such as people's personal phones, watches, and wireless headphones.

Step 3: Gather the Data

The scanner will collect data from the wireless company. Most individuals will find it incomprehensible if the information is not presented in a clear, intelligible, and structured way that highlights the organization's overall risks.

The paper should clearly state what the risk is and where it exists. I should include the SSIDs and Bluetooth discovered (if it's part of the scan).

Some testing may include:

Have the point of contact sign in to the wireless while doing the scan.

- Use the guest account for scanning.
- Connect wirelessly without credentials.
- Pair with Bluetooth devices.
- Set up a rogue access point to test for detection (organizational authority required).
- Connect to the secure Wi-Fi from outside the network.

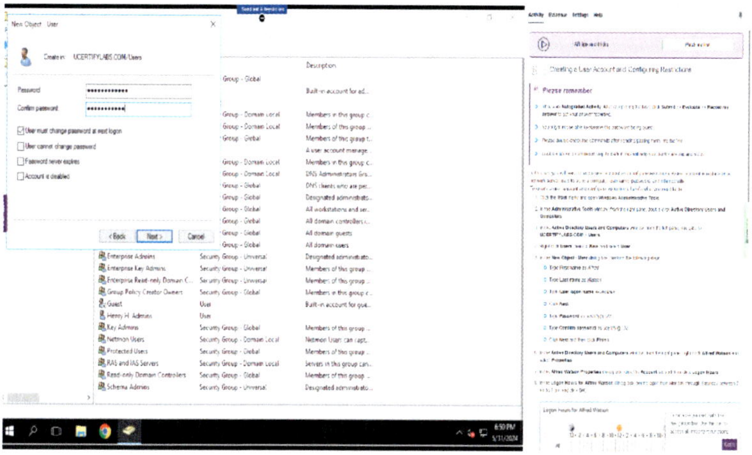

Interview of Wireless Controls

The SCA will consult with the IT personnel in charge of wireless maintenance, as well as anybody else in the organization who understands how it is configured. This might comprise ISSO cybersecurity specialists, network administrators, network engineers, or facilities managers.

The sorts of inquiries might include the following:

- How do corporate users access WiFi?
- How can visitors access the WIET?
- Does the visitor network differ from the employee wireless?
- Does the organization have rogue access detection?

• What kind of encryption is used?

• Does the organization have any Bluetooth restrictions?

Red Flags of Wireless

A totally open wireless network would be a clear red signal for both the wireless and the overall network. If you uncover one, you must wonder if the firm allows direct access from that open network to internally secured assets. An open wireless network will have no access control whatsoever.

Using outmoded and compromised encryption, such as WEP, is also a warning sign. WEP stands for Wired Equivalency Privacy. WEP is readily broken, allowing attackers to access any data on the network.

Green Flags of Wireless

If the organization has a wireless network with the highest level of encryption and authentication, it means that someone there knows about threats to the wireless.

Examination of Remote Access Controls

If remote access is permitted, the organization should clarify that it allows remote access to the system under review. Policy documentation is often high-level, leaving more technical specifics for engineering and procedural papers. Look for configuration and connection requirements, as well as implementation recommendations for each type of remote access (if there are any). You're seeking for encryption with virtual private networks (VPN).

The SCA will verify that the remote access documentation reflects the organization's current activities. This can be included in the overall security policy or in a separate remote management document approved by top management.

Interview for Remote Access Controls

In my experience, remote access evaluations are often conducted through interviews, observations of scanned material, and documentation. The SCA should speak with someone who understands and uses the existing remote access technologies. Preferably, a technical individual or group that will know things like the encryption and vendor, rather than only Here are some recommended interview questions:

- Does the system support remote access?
- What product is utilized for remote access? (This inquiry is to assess if a remote access solution is still supported.)
- What encryption is utilized for remote access? Does it employ a virtual private network?
- Who can have remote access? All users?
- Can normal and privileged users have remote access?

We are asking questions to determine whether they have a procedure in place to limit remote access and if they are enforcing the constraints that have been specified.

Testing of Remote Access Controls:

For testing, the SCA will determine if the organizational users access the system remotely, as defined in the security policy. At this step in the process, the emphasis is on the method for remote access. Here are some alternatives for testing:

- Allow the point of contact to remotely access the organizational system under testing.
- Remotely join the system with a privileged account (e.g. Gf).
- After signing in remotely, keep the point of contact logged in to check if the connection will automatically terminate after the specified period in the instructions.
- Utilize several remote access techniques for the point of contact (if the firm has many options).
- Have the system administrator log in to the remote management system and demonstrate its configuration.
- Log in remotely using a guest account.

One of the challenges with evaluating a remote connection is that the system may be designed such that you cannot test it while on-site. You would have to test it from outside the company, and assessments are often conducted on-site, thus observation and interviews are utilized instead of testing.

Red Flags of Remote Access:

This is not ideal if the company utilizes out-of-date software for remote access. There are far too many known remote access exploits targeting EOL software. If the company employs poor or (worse) no encryption, it may have already been compromised. The document demonstrates the gravity of this massive error.

Green Flags of Remote Access:

An organization that correctly implements remote access will employ not just vendor-supported solutions, but also military-grade encryption such as ABS and end-to-end VPNs. The organization will impose tight restrictions on remote access, such as timeouts for idle connections or restrictions on which systems are accessible remotely. Some firms even limit who may access which systems remotely and at what hours of day or night.

Examination of Access to Mobile Devices

AC controls feature the ability to regulate access via a mobile device. The SCA is searching for documentation outlining connection requirements as well as connection and implementation instructions for mobile device system access.

A Mobile Device Management (MDM) policy may comprise the following:

- Laptops and notebook PCs.
- Compatible with smartphones (iPhone and Android), tablets, and portable media players.

Please keep in mind that laptops and notebook computers are occasionally utilized as components of stationary workstations rather than as fully functional mobile devices. As a result, notepads and laptops were included in the MDM on an individual basis.

The MDM policy will apply to workers, visitors, and contractors who use mobile devices to access the impacted system.

This only applies to systems that allow mobile devices to access the system under review. As an example, suppose:

There are three types of mobile device usage:

The BYOD policy allows people to bring and use their mobile devices at the firm.

- Corporation-owned/business-only (COBO) devices are owned by the corporation and used solely for business purposes.
- Company-owned / personally enabled (COPE) means the gadget is owned by the company but may also be used for personal purposes.

Whatever categories are chosen, they should be recorded, with any limits stated. Some limits might include things like:

- Encrypt data at rest and in transit on mobile devices.
- Lockdown application configurations, limits, system profile settings, cloud store restrictions, device function lockdowns, device supervision, and device location monitoring.

Interview of Access for Mobile Devices

The SCA's questioning will focus on the organization's MDM policies and practices.

- Which department manages mobile device restrictions?
- Who can access the system via mobile devices?
- Which systems may be accessed via mobile devices?
- What mobile devices do you use to access the system?
- What can be done on the system via mobile device?
- Is data on the mobile device encrypted?

Testing of Access for Mobile Devices

When the SCA investigates utilizing mobile devices to access the evaluated system, they will compare the access to what is specified in the security policy. A few tests can be used to assess if the organization has restrictions:

- Allow users to access the system using a BYOD phone (according to policy limits).
- Use unauthorized mobile devices to access the system. • Ensure data is encrypted at rest or in transit. • Download random games on the mobile device.
- Check the device's management settings.

Essentially, the SCA may go through the list of limits and criteria mentioned in the documentation and test the smartphone.

Red Flags of Mobile Devices

I worked for an organization that used mobile devices on their network, but they had little or no control over the VIPs' devices. Worse, they had no idea what was on those phones or what authorized users did with them.

Mormon Hacker: Church of Latter-Day Hacks

"So, I put up a free Wi-Fi access point. I placed it outside in public areas." My hacker friend's face lit up with excitement. "You'd be astonished how many people register for it. And this gadget can collect information on all Wi-Fi networks in the neighbourhood."

"Why do you do it?" I asked him. We were both part of the local Information System Security Association. He resembled a Mormon. He wore pants and a button-up long-sleeved shirt. He had the appearance of a youthful politician that the majority would vote for without understanding his attitude. However, he shared the values of the cyber organization Anonymous.

He was so anti-establishment that I was convinced he had a safe home in the middle of the Rocky Mountains and maybe some computers rigged to explode if he needed to remove proof of illegal access to various nations. I didn't want to know, but I asked, "Do you hack their accounts?"

"No," he said. "I just do it for fun."

I have no means of knowing whether what he said was true.

This dude is on LinkedIn. I will not publish his identity for safety reasons, but you may have encountered his name in searches for pen testing or hacking. He specializes in hacking and is one of three extremely skilled cybersecurity professionals I know.

Who, should they so choose, maybe super villains. They have not gone wicked, by the grace of a higher power. In 2007, this hacker guy was involved in what is now called a "Wi-Fi pineapple." It is a method for automating account access and data collection. It is now widely known how to gain access using this approach. Anyone may purchase the kit online and install one anywhere—in a store, on a foundation, in a bank, etc.

This is only one of the many justifications for the significance of user education and access limits. SCAs need to conduct tests to make sure that the company has procedures in place for managing accounts, wireless access, remote access, and mobile devices.

Chapter 9

Awareness and Training Control Assessment (AT)

We need to find proof that the organization has a training program strong enough to fit the size and complexity of the network or system we are assessing. Awareness and Training (AT) controls cover security and awareness training for all business or mission system users.

We have pictures of computer-based testing, emails displaying recent training activities, and a spreadsheet that breaks down the training timetables. Ensuring the training is appropriate for the setting is our goal. For instance, conducting, tracking, and recording all training by email once a year might be sufficient for a local area network consisting of five computers and three active users. However, it would be absurd to manually manage and document yearly security awareness training by email for a company with 1,500 users and 400 mixed-mode systems (Linux and Microsoft servers). They couldn't possibly keep track of all the training by email alone.

The AT control family has a wide range of applications:

- Privacy

- Insider threats
- Cybersecurity
- Phishing

As assessors, we check for documented processes, the appropriateness of the training for the environment, any required specialty training, and role-based training. The following safeguards must be in place:

AT-1: Policy and Procedures

AT-2: Literacy and Awareness Training

AT-3: Role-Based Training

These controls concentrate on the training process, though businesses will still be subject to other AT controls.

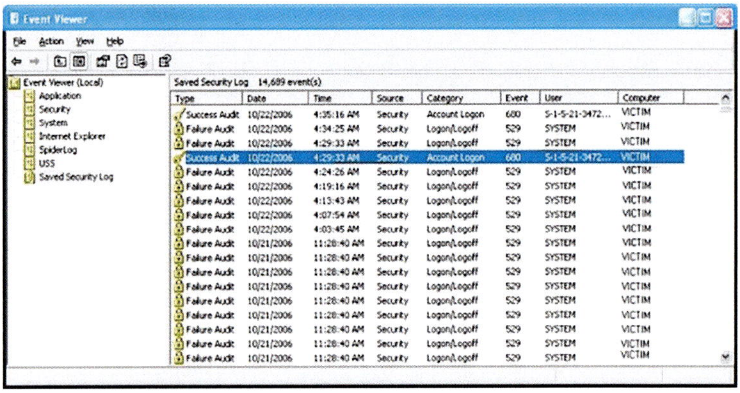

Essential Education Evaluation

Certain businesses mandate specific training. For instance, HIPAA training may be necessary for systems in the healthcare sector. Organizations with a military or intelligence focus may need training to handle sensitive material. Retail and customer service personnel might require training on PCI DSS credit card data security or consumer privacy data protection.

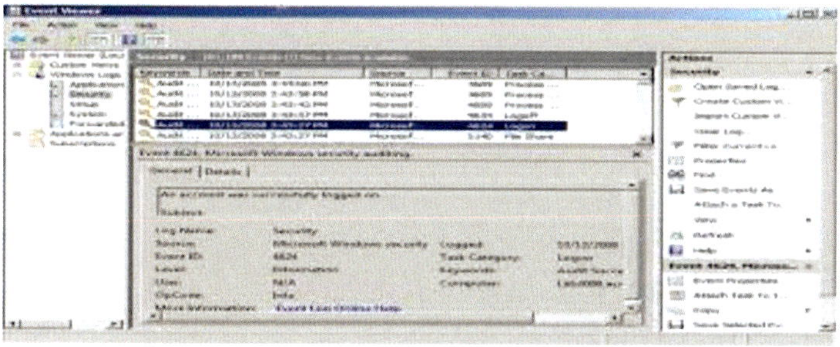

Additional specialized education might consist of:

- Phishing Campaigns
- Privacy
- Insider threat
- Records Management training (law enforcement)
- Financial data protection (financial sector)
- Government Compliance training (government)

The SCA is required by policy to consider any specific training that may exist. Check if this training is being conducted once it is noted in the policy.

Examination of Security Awareness Training

An organization should have a document outlining the security awareness training program, how often it is offered, and how it is handled. This might be a series of training manuals together with a policy, training strategy, process, or even all of these.

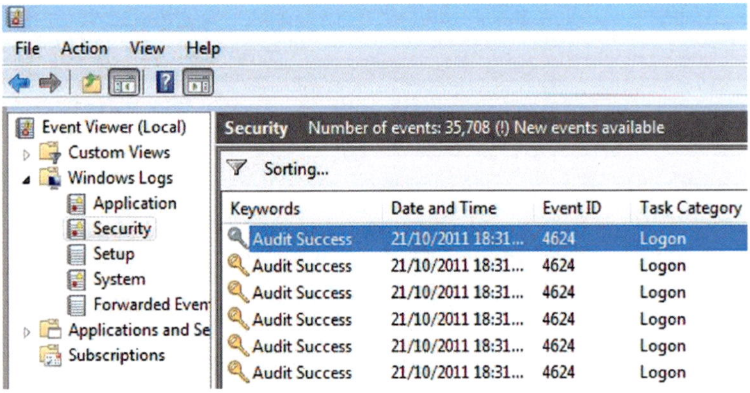

The assessor ensures that the security awareness training is highlighted in a comprehensive document. Upon reviewing this policy or strategy, they will determine if the organization has evaluated it within the past year. Additionally, if there are any current training documents demonstrating that training is conducted annually, these must be reviewed. At the very least, training should teach users how to perform fundamental security procedures.

When dealing with large, high-impact systems that demand many administrators, the assessor will review the training logs for each job.

Among the permitted artifacts are the following:

- Screenshots of completed training
- Spreadsheets of users' training schedules
- Emails of training completion
- Training certificates

All that is required of the assessor is confirmation that the organization conducts annual basic security awareness training. Additionally, training ought to be established for specific roles. For instance, cybersecurity specialists will require training appropriate to their position. Security awareness training must be accompanied by role-based training.

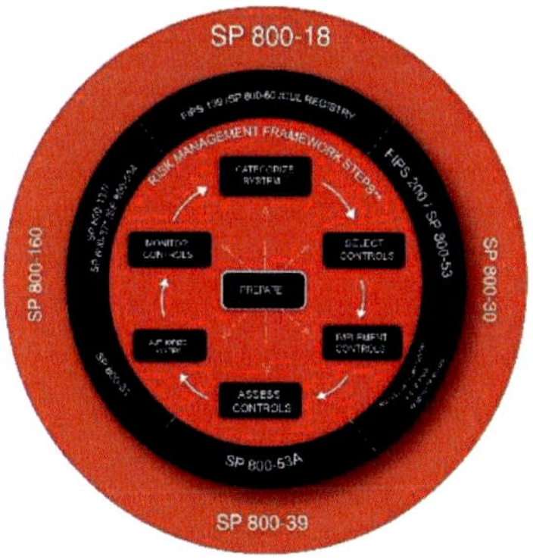

Interview for Awareness and Training

To find out if the company has an ongoing training program, the SCA may speak with system users in addition to reviewing training records and policies.

They may question a typical user in an interview with the following:

- What type of training do you receive?
- Is there security awareness training in the organization?
- When was the last time you did cybersecurity training?
- Can I see evidence of your training?
- For privileged users or special access users, the SCA might ask the following to determine role-based training compliance:
- How do you fit within the system? An administrator of the system?
- Are you trained for this role?
- How recently did you get training for those administrator responsibilities?

Red Flags of AT Controls

During interviews, the assessor notices a deficiency in training. The assessor must request training records if the subject matter expert on the system lacks a working understanding of the system to which they have been assigned. It is possible that the assessor is just speaking with the incorrect person (this occurs frequently in organizations, with many employees being unaware of this). However, if they have discussed the systems and procedures they support, there's a significant likelihood the training is insufficient. Additional warning signs:

- No training records at all
- No indication of a training site or training material
- Green Flags of AT Controls

Subject matter specialists with extensive expertise who have completed many evaluations will exude confidence. It will seem as though they practiced. They will be knowledgeable about the present procedures and the rationale behind them in addition to the system to which they have been assigned. Effective organizational training teaches the local procedure to safeguard the system since it is in-house developed rather than merely generic. They will get role-based training and instruction on basic security awareness.

A Practical Scenario: "I Blame the Army"

"Airman Smith, what time does the evaluation begin?" I didn't look away from the television. I was about finished evaluating my troop. I had until the end of the business day to complete it. I was awaiting the arrival of the evaluation team in the meantime so that I could accompany them to the location.

"Have they arrived yet? Must we collect them at the gate?"

"Sir, the assessment has already begun." Smith shook his head.

"Indeed, have you received this email? It appears that they have already joined the network. There is a breach here."

Not even my work was saved. I leaped from my seat and went to see his PC at his desk. The security operations center notified all IT managers of a compromise and the commencement of the Red Team drill through a notice.

The evaluation team was hacking our network to test it.

They had received permission to carry out the penetration test.

They were quickly and deeply infiltrating us. I just didn't know how else to put it. We were screwed, and I wanted to know just how bad it was because I needed to inform Master Sergeant Tension of this.

I contacted the security operations center contact I had made.

"Staff Sergeant Matheson, Incident Handling Security Ops, may I help you, sir or ma'am?" a harsh, deep voice said as she answered the phone. I inhaled deeply and said, "Sergeant Matheson, this is Sergeant Brown at the Information Management Office. I noticed that the assessment team received a breach notification from the

SOC just now. Could you please elaborate a little? How did they enter?"

"Phishing attack," Sergeant Matheson chuckled.

"A phishing attempt?" I cursed. "Who the heck hit that darn link? Everyone has received instruction and training. Every other month, we distribute exams and instruction. Which unit?"

Matheson replied, "Everyone has received training, with the exception of the Army. I suppose that an Army helicopter unit out of Bragg is exempt from our training requirements."

"The armed forces." I exhaled. "Stop talking now."

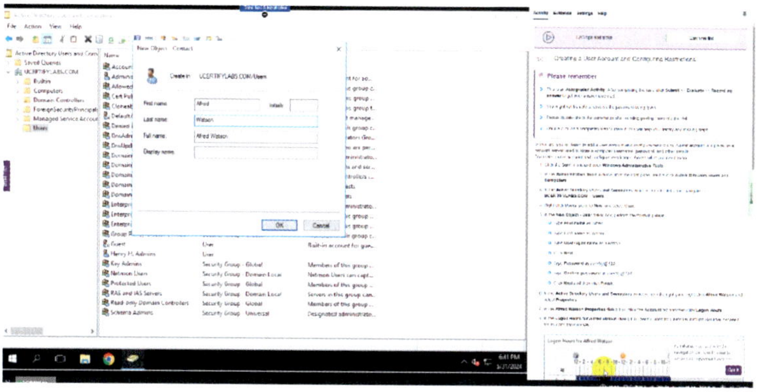

Audit and Accountability Control Assessment (AU)

For the organization's continuous maintenance, troubleshooting, and cybersecurity, audit logs are essential. The text files from all systems, including PCs, laptops, routers, software, and servers, make up the audit logs. These logs provide an overview of the system's current state. System administrators can track when and why a server restarts by looking at the logs that are generated. If the collaborative software suite malfunctions, the help desk technician gathers and examines error logs to ascertain the cause of the malfunction.

The audit logs are a useful tool for network engineers and cybersecurity experts to identify potential denial-of-service attacks when the web server and routers are operating slowly.

Accountability, audit, and AU controls ensure that event logs are submitted, appropriately tagged, and periodically examined.

We are concentrating on a few of these AU controls as assessors:

- **AU-1:** Policy and Procedures
- **AU-2:** Event Logging
- **AU-3:** Content of Audit Records
- **AU-6:** Audit Record Review, Analysis, and Reporting
- **AU-9:** Protection of Audit Logs
- **AU-11:** Record Retention

By ensuring these controls are in place and functioning properly, we help maintain the integrity, availability, and security of the organization's systems.

Chapter 10

Audit and Accountability Control Assessment (AU)

Examination of Audit and Accountability

Organizations should document system logging, with policies and procedures specifying what is logged. It would be extremely challenging to handle and keep track of all logs from all systems; thus, logging everything is impractical. Instead, refer to the documentation outlining the types of logs collected and the systems from which they are gathered.

As an SCA, it is not our job to advise on what to log. Our responsibility is to ensure the organization has specified its needs and verified the necessity for logging. This includes the accountability policy, audit, and AU-1.

We need to ensure they are collecting the data they claim to be tracking. If their policy states, "Organization XYZ ensures all failed login attempts are logged," we must find evidence of this in their system's event logs.

Recent audit logs will inform the SCA that the company has audit logs satisfying its event log criteria, addressing AU-2.

Depending on the depth of the evaluation, the SCA may examine logs from various systems in the evaluated environment. In practice, both during assessments and evaluations, the SCA typically examines business system and primary mission logs. They do not focus on logs related to Microsoft Office, printers, or other peripherals; instead, they concentrate on core functions. Nonetheless, user logs from endpoint devices can be invaluable in identifying and resolving cyber-attacks.

One way to evaluate record retention is to examine the location of log collection. Logs are typically stored on systems' local hard drives, including PCs, servers, mobile devices, and networking equipment like switches, firewalls, and routers. Organizations usually forward all their logs to a centrally maintained location. The SCA should examine these logs and search for current timestamps.

Interview for Accountability and Audit

To determine if audit logs are being reviewed, the SCA can speak with the security operations center in addition to making inspections. The assessor must consult the appropriate individuals, such as the information system security officer, cybersecurity analyst, or security operations center analyst. Relevant questions include:

- Does the organization collect audit logs?
- Where are audit logs analyzed?
- Where are audit logs stored?

- How often are audit logs examined?
- What types of events are logged?
- What systems are collecting events?
- How long are logs kept?
- The SCA can also inquire about incident response (IR) controls at this time:
- What steps does the company take in the event of a security breach?
- What kinds of security breaches have been found?
- What steps does the company take if malware or an attack is discovered?

Testing of Audit and Accountability

Enabling and gathering logs is only half of the requirements for audit and accountability. The organization should frequently review the logs.

Administrators and SCAs will collaborate to ensure logs are enabled, safeguarded, and examined. Standard users should not be able to alter event log configurations; they should only be able to access audit logs.

If an ordinary user can stop event logs, an insider threat could use this access to corrupt or modify the system and remove evidence of malicious activities.

To verify that audit logs meet the organization's criteria, the SCA should assign a privileged user to access and test the logs' settings.

A typical user might also attempt to open and change log settings. This will confirm that audit logs are protected according to AU-9 protection standards.

Event logs occasionally include personally identifiable information (PII), such as names, phone numbers, and addresses. The assessor should check or validate data protection if PII is being gathered. Protected logs should only be accessible and modifiable by authorized users.

Red Flags of AU Controls

There may be audit log content available for the organization even when the logs are deactivated. This is concerning as they have adopted the audit log features, even though they are aware of "what they are supposed to do and what the assessor wants to see."

Another problem is that the company runs two separate operating systems in a mixed-mode environment. Occasionally, they will efficiently gather logs for one but not the other, or they will have logs just for servers and nothing else. It is our duty as assessors to document what we observe.

Green Flags of AU Controls

A centralized security information and event management system that gathers all relevant logs on the systems indicates strength. It usually means they have carefully chosen the logs, are processing them, and have a team dedicated to monitoring, assessing, and reacting to the data. Small businesses or isolated systems may lack centralized security event management. It will be sufficient to have a documented procedure where they regularly check logs as determined by the organization.

Control Assessment (CA)

Evaluating the system contributes to validating the CA controls. The organization's authorization, monitoring, and evaluation procedures are covered by CA controls. Authorization and monitoring processes still need to be evaluated.

The assessor can gain a decent understanding of the organization's general compliance with CA control by concentrating on the following controls:

- **CA-1:** Policy and Procedures
- **CA-2:** Control Assessments
- **CA-5:** Plan of Action and Milestones
- **CA-7:** Continuous Monitoring

Chapter 11

Examination of CA - Assessment, Authorization, and Monitoring Controls

The SCA must note the policy's alignment with the security framework in use. While NIST 800 is used by federal systems, other cybersecurity frameworks such as CIS, ISO 27001, PCI/DSS, or others may also be used by some systems. Adherence to a security framework establishes the necessity of evaluations, approval, and ongoing observation.

The following records can be examined by the SCA to evaluate overall compliance with CA controls:

- Last security assessment report
- Last authorization package
- Continuous monitoring of policy process/procedure
- Current Plan of Action and Milestones (POA&M)

By ensuring these controls are in place and functioning properly, we help maintain the integrity, availability, and security of the organization's systems.

Are Security Assessments Being Done?

The assessor is responsible for keeping track of the date of the most recent security evaluation. The SCA should verify that security assessments have been completed within the allotted period, whether quarterly or annual, as required by the organization. The ability of the organization to conduct assessments is proof that it has a CA-2, or control assessment, procedure in place.

Is There an Authorization Process?

The organization's authorization procedure is reflected in the most current authorization package. Some organizations may not even be aware of what this process entails. The SCA should ensure that this procedure is documented in a comprehensive plan or policy statement. Additionally, the authorization package must be reviewed and approved by higher-level management, often a CIO or CISO. In the context of the NIST risk management framework, this role is known as an Authorization Official (AO).

Ways to Observe Continuous Monitoring

Continuous monitoring involves various techniques to oversee every aspect of system security, including physical security, personnel, network condition, and documentation security. To ascertain ongoing surveillance, the SCA can examine the following:

- Records of recent physical security inspections.
- Activities of the security operations center (if any).
- Actions taken to manage vulnerabilities.
- Dates of the most recent document reviews.
- Completion date of the most recent system security strategy.

The SCA must monitor the frequency of security control checks within the organization. Any vulnerabilities that cannot be repaired within the appropriate time limits should be tracked through the Plan of Action and Milestones (POA&M) procedure. Even if they do not use the term "POA&M," the assessor should look for any procedure where long-term vulnerabilities are monitored and addressed incrementally. Ignorance of vulnerabilities and lacking a plan to address them is detrimental to an organization, and attempting to hide them is even worse. Organizations might keep significant risks and vulnerabilities concealed, hoping the assessor will not notice. This Behavior suggests that they are taking significant chances and hoping they are not compromised.

As an assessor, it is crucial not to be offended if you discover that the company is concealing issues. Instead, focus on presenting the data efficiently, allowing the data to demonstrate the potential for hacking so it speaks for itself.

CA - Assessment, Authorization, and Monitoring Interview

The SCA can interview the specialists in charge of monitoring and assessment. These inquiries will focus on the entire authorization, internal evaluation, and continuous monitoring processes. The interview should be conducted with an information system security officer, compliance officer, subject matter expert, or someone with similar knowledge, as most organizational positions will not have detailed insights into these processes.

Questionnaires for the Evaluation Process (CA-2) During Interviews:

- How frequently are system assessments conducted?
- How do you determine the assessment's scope?
- What types of system evaluations are conducted?
- Who is in charge of the evaluations?
- How often are impartial evaluations carried out? (CA-2)
- Authorization and POA&M Interview Questions (CA-5, CA-6):
- When was the last time the CIO or authorized officer approved the system?
- What is the process for obtaining authorization for this system?
- Who approves the use of this system?
- Does the system have any open plans of action or milestones?

- What is the POA&M procedure?

The SCA compares the organization's actions to the security policy's provisions on authorization, assessments, and continuous monitoring. They also assess whether the company genuinely has these procedures in place and how reliable they are. Sometimes, the company might have procedures for authorization, continuous monitoring, and evaluation, but some activities might be absent for the system currently undergoing assessment.

Testing of CA — Assessment, Authorization, and Monitoring

Assessors can have operators or system administrators demonstrate their monitoring procedures, allowing SCAs to evaluate how the organization conducts assessments and monitoring. Assessors might also ask the organization to show how endpoint devices are detected if they use automated monitoring technology.

The SCA might look for answers to the following:

- When is monitoring completed?
- Is the individual filling the position qualified to do it?
- How exhaustive is the monitoring method?

The internal assessment procedure would be tested similarly. The subject matter expert who conducts routine scanning will meet with the assessor and run the internal scanning tool.

Red Flags of CA Controls

If systems are not routinely assessed, oddities impacting mission and key business operations may occur. Some observations include:

- Numerous systems nearing the end of life are still online.
- Computers running unpatched software for years.
- Unidentified servers.
- Equipment that should have been retired months ago.
- Unapproved or unsupported software.

Green Flags of CA Controls

Organizations that regularly assess their risks and have management approve the findings typically have less overall risk. These organizations usually have a large number of recent evaluations.

CM - Configuration Management Control Assessment

The SCA looks for whether the organization has a program managing settings and system modifications to evaluate configuration management. The assessor must determine if impact analysis techniques are used and whether the principle of least functionality is applied. The organization should have strict control over the configuration and appearance of its systems.

If the organization has documented network-wide configuration management concerns, it indicates an attempt to mitigate risks. A lack of configuration management indicates a serious systemic failure that impacts other areas of the organization's security posture. If these concerns are not addressed, it is a significant red flag.

Some areas for the assessor to focus on include:

- **CM-1:** Policy and Procedures
- **CM-2:** Baseline Configuration
- **CM-3:** Change Control
- **CM-4:** Impact Analysis

Chapter 12

Contingency Plan Control Assessment

Assessing whether an organization can withstand natural catastrophes, man-made disasters, or any other likely hazards depends on the contingency plan (CP) controls in place. During an assessment, an assessor must find proof that the company can carry out its core operations in the event of a system failure or, at the very least, that it has a backup plan in place. The primary controls for CP to evaluate are:

1. **CP-1: Policy and Procedures**
2. **CP-2: Contingency Plan**
3. **CP-3: Contingency Training**
4. **CP-9: System Backup**

Examination of the Contingency Plan Controls

If the organization has a contingency plan, it should include several key documents. A comprehensive contingency plan document should be in place, covering every system under the department's jurisdiction. This document requires senior management's approval and signature, proving that they take CP controls seriously. Another component of CP controls is backups. A policy should require the adoption of a backup procedure for vital systems.

Underlying this high-level CP strategy should be specific processes or guidelines outlining actions in the event of certain calamities. For instance, there may be protocols for a system breakdown or a natural disaster.

The assessor needs to note the revision date on the document. However, when it was utilized is more significant than when it was reviewed. A common poor practice is creating a contingency plan policy that an organization never utilizes.

The company is expected to prepare and execute the backup plans before recording the meeting. They should conduct a "tabletop" or simulated disaster exercise to review their actions in a crisis scenario. An after-action report documents this exercise.

The assessor can verify if the company has a contingency plan by looking at several CP artifacts, such as:

- The system's continuity, disaster recovery, and business continuity plans
- Proof of Disaster Recovery/Continuity Plan training, including:
- Reports on actions taken afterward

- Remarks about the tabletop task
- Proof of backup strategies, including screenshots and backup policies and procedures
- Additional plans to consider, such as:
- Plans for Continuity of Operations
- Response plans for breaches
- Critical Infrastructure plans
- Insider threat plans

Interview for CP - Contingency Plan Controls

During the interview, focus on whether the organization has employed the contingency plan, updated the disaster recovery and backup plans, and performed actions to restore the backup and recovery systems following a disaster. The assessor may inquire about the following CP topics:

- Do you have a method for backups?
- Which systems have backups?
- Can I view the settings for your system backup?
- Have you tried the recovery process for your backup?
- Do you have a disaster recovery or contingency strategy in place?

- How recently have you needed to apply the disaster recovery process?
- Have you utilized disaster recovery or contingency plans during practice?
- Can I view your notes or reports on the actions taken during the most recent contingency plan exercise?

Red Flags of CP Controls

A red flag is when there is no proof that the CP controls are tested, exercised, or carried out in any way. The assessor should follow up with more inquiries and documentation. Most organizations will have some form of business continuity or disaster recovery plan addressing contingency planning

Green Flags of CP Controls

It is encouraging when a company has tested its business continuity, disaster recovery, and continuity of operations within the previous year and has documented the findings.

Chapter 13

Contingency Plan Control Assessment

Assessing whether an organization can withstand natural catastrophes, man-made disasters, or any other likely hazards depends on the contingency plan (CP) controls in place. During an assessment, an assessor must find proof that the company can carry out its core operations in the event of a system failure or, at the very least, has a backup plan in place. The primary CP controls to evaluate are:

1. **CP-1: Policy and Procedures**
2. **CP-2: Contingency Plan**
3. **CP-3: Contingency Training**
4. **CP-9: System Backup**

Examination of the Contingency Plan Controls

If the organization has a contingency plan, it should include several key documents. A comprehensive contingency plan document should be in place, covering every system under the department's jurisdiction. This document requires senior management's approval and signature, proving that they take CP controls seriously. Another component of CP controls is backups, with a policy requiring the adoption of a backup procedure for vital systems.

1. Underlying this high-level CP strategy should be specific processes or guidelines outlining actions in the event of certain calamities. For instance, there may be protocols for a system breakdown or a natural disaster.

2. The assessor needs to note the revision date on the document, but when it was utilized is more significant than when it was reviewed. A common poor practice is creating a contingency plan policy that an organization never utilizes.

3. Thc company is expected to prepare and execute the backup plans before recording the meeting. They should conduct a "tabletop" or simulated disaster exercise to review their actions in a crisis scenario. An after-action report documents this exercise.

The assessor can verify if the company has a contingency plan by looking at several CP artifacts, such as:

- The system's continuity, disaster recovery, and business continuity plans

- Proof of Disaster Recovery/Continuity Plan training, including:
- Reports on actions taken afterward
- Remarks about the tabletop task
- Proof of backup strategies, including screenshots and backup policies and procedures
- Additional plans to consider, such as:
- Plans for Continuity of Operations
- Response plans for breaches
- Critical Infrastructure plans
- Insider threat plans

Interview for CP - Contingency Plan Controls

During the interview, focus on whether the organization has employed the contingency plan, updated the disaster recovery and backup plans, and performed actions to restore the backup and recovery systems following a disaster. The assessor may inquire about the following CP topics:

- Do you have a method for backups?
- Which systems have backups?
- Can I view the settings for your system backup?
- Have you tried the recovery process for your backup?

- Do you have a disaster recovery or contingency strategy in place?
- How recently have you needed to apply the disaster recovery process?
- Have you utilized disaster recovery or contingency plans during practice?
- Can I view your notes or reports on the actions taken during the most recent contingency plan exercise?

Red Flags of CP Controls

A red flag is when there is no proof that the CP controls are tested, exercised, or carried out in any way. The assessor should follow up with more inquiries and documentation. Most organizations will have some form of business continuity or disaster recovery plan addressing contingency planning.

Green Flags of CP Controls

It is encouraging when a company has tested its business continuity, disaster recovery, and continuity of operations within the previous year and has documented the findings.

Chapter 14

Identification and Authentication Control Assessment

Assessors examine the processes and procedures related to multifactor authentication (MFA), passwords, and identity and authentication (IA). They evaluate if the company has established MFA, password, or username rules and processes and if these guidelines are followed as documented.

The most effective method for evaluating how the system uses authentication and identification rules is to combine testing and observation. Most of the information required by the Security Control Assessor (SCA) will be provided by observing the differences between the IA controls applied in the system's baseline configuration and those specified in the security policy and procedures. During an evaluation, any number of controls may come up, but the primary IA family controls to validate are:

1. **IA-1: Policy and Procedures**
2. **IA-2: Identification and Authentication (Organizational Users)**
3. **IA-4: Identification Management**
4. **IA-5: Authentication Management**

The SCA can check the controls' implementation using a variety of techniques. They may test the login procedure, monitor the IA controls, and speak with the stakeholders who oversee the IA process.

Analyzing Identity and Authentication (IA) Controls

The company needs to have a policy covering the process of identification and authentication, including MFA, two-factor authentication (2FA), and constraints on usernames and passwords. The documented security safeguards should meet the minimal constraints specified by the industry.

For instance, if the business is part of the federal government and mandates a 20-character password, but the organization only requires 16, this discrepancy should be mentioned in the security assessment report. Additionally, if the IA policy specifies exclusive use of MFA, any system requiring just a username and password must be documented.

To conduct the evaluation, the SCA can look for the following signs:

- A documented baseline with the password and username settings

- The system's username and password settings
- System multifactor scttings
- Policy for identification and authentication
- Procedures for identification and authentication

Interview for IA - Identification and Authentication

The SCA performs this test to ensure the system's actions align with the approved policy. They have access to user privileges, mandatory banners, audit log settings for unsuccessful login attempts, user session limits, and other security measures required for the login procedure. The assessor may inquire about the following IA topics:

1. Do the systems employ multifactor authentication or username/password?

- Asking the appropriate subject matter experts can reveal whether various systems employ distinct login procedures.

2. What is the password complexity?

- The SCA should look for the password's length, use of capital and lowercase letters, digits, and special characters. This should match the organization's policy.

3. Explain the MFA procedure.

- Variations of MFA include something you own (RSA fob, smart card), something you are (facial recognition, fingerprints, retina scan), and something you know (password, PIN code).

4. How are mobile device logins made? (If relevant)

- If mobile devices are present, a different authentication procedure may be used.

5. Does logging in remotely require a separate login, password, or MFA process? (If relevant)

- Remote IA controls might differ from local login procedures.

6. Does the login procedure vary depending on the operating system? (If relevant)

- In environments with multiple operating systems, different systems might have distinct login procedures and IA controls.

Testing IA - Identification and Authentication

Subject matter experts within the company can help the SCA verify identity and authentication. These might include system administrators or cybersecurity experts who can demonstrate the system's Behavior during the login procedure. The SCA or subject matter expert can carry out the following tasks to test IA:

I. Attempt to log in using fictitious or non-existent credentials.

- This checks if the system has proper identity and authentication measures and monitors what happens during an unsuccessful login attempt (IA-2 and IA-3).

II. Sign in using a regular user account and try to access administrative tools.

- This examines access controls (AC). Each user should have access limits based on their responsibilities after identification and authentication. Regular users should not have the same access as privileged users (AC-3).

III. Request that the identity and authentication manager be accessed by the system administrator.

- This verifies settings at IA-2, IA-3, IA-4, and IA-5 if the company has a centralized way to manage the identity and authentication process.

Red Flags of Identification and Authentication

Red flags include:

- Absence of multifactor authentication (unless the system cannot support it).

- Shared passwords and usernames (especially for privileged accounts).
- Systems without login and password requirements, enabling weak passwords.
- No standard for username naming conventions.

Green Flags of Identification and Authentication

Green flags include:

- An organization moving towards universal MFA integration.
- Minimal use of usernames and passwords on the assessed system.

Chapter 15

Incident Response-IR

The Security Control Assessor (SCA) must determine if the company has a procedure in place for Incident Response (IR) controls that address the system under evaluation. While most organizations have security policies that address incident response, the assessor's focus should be on the system they are evaluating. Does this system fall under the purview of the current incident control documentation?

If an event occurs in the target system, the assessor must confirm that a strategy and procedure are in place for handling it. These systems are often covered under the organization's overall incident response strategy. The incident response procedure for the system under evaluation should include:

- ***Preparation:*** The Company identifies which attacks are most likely to exploit known vulnerabilities.
- ***Detection & Analysis*:** When a security event is discovered, the company assesses its impact.
- ***Eradication & Containment*:** The Company addresses and contains the security issue.

- ***Post-Incident Activities*:** The Company learns from the incident, takes preventive measures, and prepares for future occurrences.

The following controls are particularly beneficial to the SCA:

- **IR-1:** Policy and Procedures
- **IR-2:** Incident Response Training
- **IR-4:** Incident Handling
- **IR-5:** Incident Monitoring
- **IR-6:** Incident Reporting
- **IR-8:** Incident Response Plan
- Examination of Incident Response Controls

The documentation of the incident response procedure must be thorough, providing proof of an operational incident response plan. These documents should include both a comprehensive guide on addressing specific incidents and an overarching strategy for the IR efforts. The assessor should look for the following:

***Incident Response Strategy or Plan*:** These high-level documents are intended for a broad audience and outline the incidents covered, who is impacted, and why incident response is necessary. The assessor should verify that these records are regularly reviewed.

***Incident Response Procedures*:** These provide detailed guidance for users and stakeholders on what to do in the event of a security issue, such as how to initiate an incident response team. There may be specific procedures for various types of incidents.

***Proof of Incident Response Training*:** Minutes, notes, or after-action reports (AARs) can indicate whether training is conducted to prepare for incidents. The SCA ensures that the policy is implemented and incident procedures are applied.

***Proof of Actual Incident Response*:** Emails, reports, and tickets can serve as evidence of real incident response. These documents show how the company adhered to the IR policy and procedures during a security incident.

Interviewing for Incident Response Controls

To verify the controls, the assessor might meet with the security team and inquire about the status of the incident response. If the company routinely practices the IR process, the cybersecurity staff should be able to answer the following questions:

- Is incident response documented?
- When did the last security incident occur, and what methods were applied?
- Is there an incident response team? Who is part of it?
- Can you show me your IR records?
- Do you practice security incident response?
- How do you detect security breaches?
- Testing Incident Response Controls

The assessor may ask the involved parties to confirm if the company offers training, issue reporting, and automated testing. The demonstration should illustrate an operational process, primarily focusing on high- or moderate-impact automated systems.

Automated Testing (IR-3(1)): The company uses an automated system to assess the efficacy of the incident response.

Automated Incident Reporting (IR-6(1)): Security information and event management systems, such as Splunk and LogRhythm, can automatically report and respond to incidents.

Red Flags of Incident Response Controls:

Red flags include the absence of a comprehensive procedure for dealing with security events. If the organization lacks an overarching policy with detailed procedures for handling incidents, this indicates deficiencies in the IR and other controls.

Green Flags of Incident Response Controls:

A strong incident response procedure is demonstrated when an organization effectively manages and documents a security event. This includes having a well-documented strategy and actively practicing and refining their incident response processes.

Chapter 16

Maintenance-MA

The organization places a high value on maintaining its systems and supporting infrastructure, encompassing everything from vendor contracts to HVAC system inspections. The Security Control Assessor (SCA) needs to verify that the company conducts routine inspections and maintenance of both supportive and mission-critical systems.

Maintenance Controls

Maintenance (MA) controls include:

a. Maintaining and replacing outdated systems
b. Tracking system maintenance notifications
c. Performing diagnostics
d. Conducting preventive upkeep

The most used and visible MA controls are:

- **MA-1:** Policy and Procedures
- **MA-2:** Controlled Maintenance
- **MA-3:** Maintenance Tools

Examination of Maintenance Controls

The SCA is responsible for verifying the validity of vendor maintenance contracts for mission-critical equipment. Even if not every system has a vendor support agreement or maintenance contract, the assessor must understand the procedure for repairing systems in the event of a failure. For example, if a power surge destroys the power supply on a server that handles the company's primary operations, there must be a documented procedure for repairing the system, which the assessor must review.

The surrounding infrastructure supporting the mission, such as heating, cooling, and power systems, also requires attention. The assessor can check the maintenance records for the generator, if required, and HVAC systems (heating, ventilation, and air conditioning). Key systems to review include:

- Generator maintenance logs
- HVAC maintenance logs
- Vendor support agreements
- Visitor records indicating vendor access

Maintenance contracts

The assessor looks for logs, agreements, and contracts from the previous year to demonstrate the organization's ongoing upkeep. Signs of inadequate maintenance quality include:

- Systems not maintained for a year

- A/C units not serviced for years
- Generators that haven't needed repair in years
- Lack of maintenance agreements

Interviewing for Maintenance Controls

The facilities manager or equivalent is one of the best people to discuss the support systems, power supply, and environmental controls. This includes site managers, groundskeepers, custodians, and anyone responsible for ensuring generators are fuelled and HVAC systems are operational. The assessor should also speak with managers, subject matter experts, or the systems themselves for system support.

- Interview Questions for System Maintenance:
- Do you have a system service agreement?
- Is the service agreement up to date?
- When was the system's last maintenance completed?
- How do you handle a malfunctioning system?
- How do you grant maintenance personnel access to secure locations?
- Interview Questions for Facility Managers:
- When was the last maintenance performed on the HVAC system?
- When was the generator's last maintenance performed?
- Is there a contract in place for generator and HVAC maintenance?

- Are there alarms for system failures?
- What happens in the event of a power outage?
- How does the company respond to an HVAC system breakdown?

Red Flags of Maintenance Controls

A lack of maintenance records is a clear indicator of inadequate MA controls. Every support system, including HVAC and generators, needs to track its service history, whether manually or automatically. It's not uncommon for evaluations to reveal sites lacking current contracts and agreements.

Green Flags of Maintenance Controls

Positive indications include a well-informed facility manager who is aware of the site's maintenance activities. Unlike system administrators, compliance officials, and cybersecurity professionals, facility managers typically have a clear understanding of the ongoing upkeep of backup systems like HVAC and generators, as they are responsible for scheduling and overseeing maintenance.

Chapter 17

Media Protection Control (MP) Assessment

The assessment of Media Protection (MP) controls involves interviews and observations to evaluate the procedures in place for safeguarding media. Media in an organization includes, but is not limited to:

- Drives
- Printouts
- Flash drives
- Discs
- Publications

Any item containing information that is not part of a computer system is considered media. The assessor's role is to determine whether this media is adequately protected. Printouts must be covered and labelled with appropriate categories to ensure the organization's media is safeguarded. Both digital and non-digital media must be used, stored, transported, and sanitized according to established procedures within the company.

Key MP Controls

Observing the top MP controls provides significant insights for the assessor:

- **MP-1:** Policy and Procedures
- **MP-2:** Media Access — controlling access to digital and non-digital media
- **MP-3:** Media Marking — marking the proper classification of digital and non-digital media

Examination of Media Protection Controls

During the assessment, the SCA looks for drives, papers, discs, and other materials left outside secure areas. The assessor will be particularly vigilant about any loose papers on workstations and printers, checking these records for sensitive information. Documents containing sensitive data, personally identifiable information, or classified information should be labelled, covered, or stored securely. If the assessor views classified documents, this must be documented, as it may violate MP regulations. If flash drives are permitted by the organization, they must be properly labelled.

Interview for Media Protection Controls

A brief visit may not reveal the full scope of the organization's media protection procedures. Therefore, interviews are essential to gather comprehensive information. The assessor needs to understand how the company covers, labels, sanitizes, and moves both digital and analog media.

Key interview questions include:

- How do you protect sensitive documents, disks, and flash drives?
- What classifications of data do you have, and how do you label them?
- Is there a policy for handling sensitive documents left out in the open?
- If flash drives are allowed, how are they labeled?
- Where are disks, drives, and documents with sensitive information stored?
- How are sensitive documents transported between sites?
- How are sensitive documents destroyed when no longer in use?
- How are hard drives and flash drives erased or sanitized for reuse?

The assessor aims to determine if the organization's media protection procedures are comprehensive enough to safeguard the sensitivity of the information. For instance, a nonprofit organization that creates and manages publicly accessible websites has different media protection needs compared to a US Air Force missile defense unit with highly sensitive materials. The variety of media permitted increases the risk of exposure.

Red Flags of MP Controls

Indicators of inadequate MP controls include finding sensitive information scattered around the building without proper coverings or security. Some companies may not have any procedures to protect critical media, posing a significant risk.

Green Flags of MP Controls

A strong indication of effective media protection is when an organization has a comprehensive media protection strategy and practice in place, far surpassing the standards of many other organizations. This demonstrates a commitment to safeguarding all forms of media and maintaining the confidentiality, integrity, and availability of sensitive information.

Chapter 18

PE - Physical and Environmental Protection

The SCA may need to inspect both the interior and exterior of the location where the system is housed if the physical controls fall within the purview of the assessment. The SCA must document the system's physical defenses, ensuring that physical barriers are in place to safeguard the system. Physical limitations and a stable environment that supports the system's fundamental operations are examples of PE controls. These support systems provide controls for humidity, temperature, and power.

Examination of PE - Physical Access Control

Organizations must determine the level of physical security required as they are in the best position to recognize their risks and weaknesses. Documentation will inform the assessor of any anticipated features, such as the presence of a fence around the area, an electronic badging procedure for authorized staff, or a physical security unit that checks badges at a front desk. The primary controls addressed in the policy are:

PE-2: *Physical Access Authorizations* - Keep a record of individuals permitted entry into the building housing the system.

PE-3: *Physical Access Control* - Implement authorizations for physical access to the system.

The level of physical security measures should correspond with the significance and categorization of the data being handled, transferred, and stored. The assessor looks for both operational and physical security measures. Physical security involves limiting or discouraging access through barriers and force, while operational security includes employee actions and policies to safeguard the system.

Not every location will require all physical security boundaries, but it is important to implement everything specified in the physical security specification. The more critical the information, the more stringent the security restrictions should be.

Observed physical controls include:

A list of authorized persons or an electronic system verifying authorization:

- Locked doorways
- Supervised parking lots
- Monitored entry points
- Security division inspections of doors
- Alarm systems for accessible windows (for high-impact systems)
- Secure windows that prevent viewing of critical data
- Fencing (for high-impact systems)
- Vehicle barriers (rocks, bollards, etc.)

Motion sensors and alarm systems

Observed operational security procedures include:

- Methods to identify authorized persons (badges, name tags, dress codes)
- Employee questioning before entering restricted areas
- Electronic badging systems requiring identification upon entry
- Metal detectors or pat-downs if necessary
- Biometric systems upon admission
- Access PIN codes
- 24/7 monitoring of mission-critical systems
- CCTV with recorded and reviewed footage
- Personnel with operational training authorized to confront potential intruders

Evaluation of Environmental Controls in PE

The assessor examines the primary systems that support the information system to evaluate the environmental controls. These include systems for water, electricity, humidity, and temperature. While each environment is unique, the primary focus is on:

Temperature & Humidity:

Extreme temperatures can damage the information system. The assessor will check if the organization has a heating, ventilation, and air conditioning (HVAC) system that:

- Accommodates the size of the system under evaluation
- Controls humidity in extremely dry conditions
- Receives routine maintenance
- Operates in a locked, restricted area

Water Damage Protection:

Organizations with significant resources in one location, such as data centers and server rooms, should consider water damage protection. Facilities often have water pipes passing through rooms with critical assets. To mitigate risks, these rooms should have master cut-off valves or isolation valves to stop water flow in case of a leak. Some computer rooms are equipped with water leak-detection systems.

Relevant PE controls include:

PE-14: Environmental Controls - Ensure that the surrounding temperature, humidity, pressure, and radiation are maintained. If required, monitors and alarms should be installed to control the environment.

PE-15: Water Damage Protection - Protect against system damage from leaks with a master shutdown feature.

Interview for PE - Physical Access and Environmental Controls

Interviews are an effective tool for evaluating operational and environmental security. Operational security focuses on the actions of authorized workers to safeguard physical security. The assessor should interview personnel familiar with both operational and physical controls, such as the security unit and facilities managers.

Questions about Physical and Operational Security:

- How are facilities accessible to authorized users? Do you use electronic cards?
- Is there a security unit on the property?

- Are there security systems in place?
- Are there cameras on the premises?
- Are cameras under surveillance?
- How long is camera footage stored?
- Is there a secure docking area? How is it protected?
- Do employees wear badges to prove they are permitted in specific areas?
- Are alarm systems installed on doors and windows?
- How frequently are alarm systems checked?
- Has there ever been an incident involving physical security? What was the organization's response?

Questions about Environmental Controls:

- Is there a backup generator? Does it have fuel on hand?
- What is the generator's lifespan? Can it continue to power necessary operations during a prolonged outage?
- Is there air conditioning? Is the HVAC unit secured? Who has access to the HVAC system?
- When was the HVAC system last maintained?
- Is the system located in a confined space with temperature controls?
- When the generator was last maintained?
- Does the server room have a master shutdown switch or isolation valve?
- Are there instructions or training programs for handling potential fire or water damage?

- Are there alerts for water detection?

Testing of PE - Physical Access Control

The best way to test alarms, locks, and procedures is to observe them in operation. Assessors can use the following methods to evaluate physical security:

- Attempt to enter doors at random.
- Try entering restricted locations without a badge or credentials.
- Trigger an alert system.
- Enter through an uncommon access point to simulate unauthorized entry.
- Use a fictitious ID to test security response. (with prior authorization)
- Sneak around the building to see if you are noticed or confronted..

Red Flags of PE Controls

High-risk systems will quickly reveal inadequate PE controls. Indicators include individuals without properly displayed badges, lack of security apparatus, and absence of cameras, raising concerns about how the system is secured.

Green Flags of PE Controls

A round-the-clock security force significantly reduces the risk associated with PE controls. Organizations with the financial means to maintain such security measures can effectively recognize and mitigate a wide range of hazards.

Chapter 19

PL – Planning

PL controls deal with system security plans, privacy plans, rules of behavior, and other essential documents. The assessor can have the organization's point of contact (POC) send the appropriate files before their arrival. The assessor will review these documents to help validate controls. The most foundational PL controls include:

- **PL-1:** Policy and Procedures
- **PL-2:** System Security and Privacy Plan
- **PL-3:** Rules of Behavior

Examination of PL - Planning Controls

The assessor will first request the system security plan (SSP). This document summarizes all security controls on the system and serves as a reference for implementing controls effectively. Having the SSP will make the overall assessment easier.

Before evaluating the username and password setup, the assessor will review the SSP's identification and authentication controls to ensure proper implementation. This allows for cross-referencing all controls in the security assessment plan to ensure they are properly documented, implemented, and applicable. The SSP may be named differently and come in various formats, such as a spreadsheet, database, content management system, PDF, or a collection of text documents. Content management systems like E-MASS, Archer, or Xacta often manage it. Sometimes organizations call it by a contract name, such as Cyber Security Assessment Management (CSAM) or Information Technology Security Risk Management. Regardless of the name, this collection of information summarizes how each control is installed and configured. These are termed "implementation statements." The assessor needs to examine statements on each security control identified in the security assessment plan.

Interview for PL - Planning Controls

As an assessor, it is essential to ask the assigned cybersecurity professional questions about their system's security policies. Double-check the documents and inquire about any missing controls to ensure accuracy. My PL questions revolve around the provided documentation, which I review and discuss while they are present.

When evaluating PL controls, consider whether there are relevant implementation statements describing how the security control is being implemented. Questions to ask include:

- Is there evidence to support the implementation statements?
- Has the system security plan been updated within the last year?
- Does the system have a security plan?

The SCA may also assess the system security strategy and consider the concept of operation (CONOPS). This section provides an overview of the system's purpose and essential functions for achieving its principal mission or aim. The system security plan typically includes an executive overview and description of the system under assessment. Additional documents may provide further information on the system's CONOPS. The system security strategy includes this and other essential PL controls.

Rules of Behavior

Organizations typically define rules of Behavior in user agreements or HR policies. The assessor can analyze these materials to identify any standards of Behavior. These rules define what users may and cannot do on the system being examined. Users must read and agree to the organization's conditions before accessing the system.

The assessor can check the following documents for the presence of rules of Behavior (PL-3):

- User permission agreement
- Human resources policy on users
- User orientation documentation
- Social media limits for users
- Signed acknowledgment
- Policy on standards of Behavior and security

Red Flags of PL Controls

A significant red flag is the absence of a system security plan. Some locations might create an overarching security plan without a detailed breakdown of controls. Without documentation, it is challenging to ensure thorough security procedures in medium and large environments.

Green Flags of PL Controls

Assessors appreciate having access to the most recent system security strategy for the system under evaluation. This document provides an overview of the system's scope, functionality, and security structure, aiding in the evaluation of controls.

Chapter 20

PM - Program Management Assessment

PM controls govern how a company manages information security and risk. Assessments typically focus on specific system controls; therefore, these controls may not always be appropriate. Implementing PM controls demonstrates to assessors that a company has security practices in place for the system under assessment. For example, PM-9 outlines the organization's risk management approach. The assessor can use any risk management technique to determine if the system has mechanisms for dealing with security events and vulnerabilities, including internal and external controls.

Examination of PM - Program Management Controls

PM controls focus on how the organization manages risk. Documentation can be helpful when conducting the assessment.

Information Security and Privacy Resources (PM-3)

This control focuses on the organization's financing for system security and privacy. The assessor can utilize the PM-3 material to determine the necessary "allocation of resources" (SA-2). The system requires proper financing and preparation for privacy and security.

Plan of Action and Milestones Process (PM-4)

PM-4 assures that an organization has a plan of action and milestones (POA&M) for all systems. If a system control is lacking, the assessor will look for a POA&M. Implementing this technique can significantly benefit a company.

System Inventory (PM-5)

System inventory is critical for baseline, configuration, and other controls. Assessors can cover several controls by examining, testing, and asking questions regarding the organization's system inventory. Validation of system inventory includes controls such as AC-3, CM-8, and PL-8.

Enterprise Architecture (PM-7)

PM-7 assures that organizations implement privacy and security aspects in system development and maintenance. The assessor will review minutes from change management or configuration meetings involving the cybersecurity team. The assessor will also review audit records for system and security applications, including malware protection solutions.

Risk Management Strategy (PM-9)

This control refers to the organization's risk management approach. Assessors should identify a risk management plan for the organization's activities, systems, and personnel. This has a broad scope, but specific outcomes may include:

- Threat detection
- Security operations center
- Vulnerability management
- Incident response reports
- Authorization Process (PM-10)

The assessor can review any updated and management-approved documents outlining the system authorization procedure. These documents can help confirm the CA-6 authorization control and outline the authorizing official, the authorized official for common controls, and the authorization procedure.

Red Flags of PM Controls

PM controls refer to how a company manages its information security program. If an organization's information system security strategy is inconsistent, it can compromise the security of the system under assessment.

Green Flags of PM Controls

When a business has a strong information system security program, risk strategy, and system inventory, it helps to keep all systems on track. The effectiveness of the strategy largely depends on how well it is implemented.

Chapter 21

PS — Personnel Security Assessment

As soon as the assessor arrives on-site, they begin analyzing PS controls and personnel security. PS controls evaluate the organization's management of individuals who have access to facilities, resources, and contact with mission-critical staff. The assessor extensively vets and determines the level of access granted to individuals within the company. The business should assign a risk rating to each position and screen candidates accordingly. For example, a housekeeping position requires a different screening level than a senior executive analyst with access to secret data. Both roles require screening, but the position with a higher degree of access will require more rigorous screening.

PS controls include policy and procedures (PS-1), personnel screening (PS-3), termination (PS-4), and access agreements (PS-6).

Examination of PS — Personnel Security Controls

The assessor should take note of how the organization screens individuals for temporary site and resource access. If there is no method for vetting an outside assessor, this is a red flag. This will prompt the assessor to investigate the documented personnel security process.

In addition to observing the procedure, the assessor can check the following personnel security documents:

- Personnel security plan
- Orientation and screening documents
- Screening policy and procedures
- System security plan
- Procedures for each position category
- Risk designations for each position
- Position descriptions

Interview for PS — Personnel Security Controls

If there is no mechanism in place to screen an outside assessor, the assessor should inquire about the lack of a personnel security process. Assessors often interview a subject matter expert on personnel security. This might be a cybersecurity representative, personnel screener, human resources manager, security manager, or any other approved individual familiar with the organization's personnel security procedures. Key questions include:

- What is the screening procedure for new employees?
- Are there various screening thresholds for different positions (e.g., background checks)?
- What is the process for departing employees and contractors?
- Is there a procedure for personnel transfers?
- The assessor is looking for confirmation that there is a screening procedure that varies based on the job role.

Red Flags of Access Management

As an assessor, it is a red flag if I can enter an organization with little or no verification of my identity and no pre-authorization. While obtaining permission to evaluate the system can be cumbersome, it is necessary.

Green Flags of Access Management

If a company is thorough and rigorous with their security protocols, they will likely apply the same standards to new organizations requiring access to mission-critical assets. Being accompanied by a temporary badge is a positive indication.

Chapter 22

PT — Personally Identifiable Information (PII) Processing and Transparency Assessment

Assessors must recognize and safeguard personally identifiable information (PII) as it is highly sensitive. They will evaluate network security, operational security, and access restrictions, as well as the protection of PII. If the assessor's role is limited to reviewing the system security plan for approval, completion, and current status, they should still be vigilant about PII. Unnecessary PII, such as contractors' full names, dates of birth, home addresses, and phone numbers, should be identified and reported. The assessor expects the company to secure systems containing sensitive personal information, including social security numbers and medical data. If the evaluation includes websites, the assessor may check if the organization allows users to consent to PII collection.

Examination of PT — Personally Identifiable Information (PII) Controls

When reviewing documents and systems, assessors should be vigilant for employee and customer information. They should check if this information is protected, publicly available, or placed in common areas. Assessors anticipate finding full names, work emails, and phone numbers, but they will also look for PII that may be used to manipulate individuals.

The assessor looks for the following PII:

- Full name, phone number, and address
- Social Security numbers
- Passports
- Driver's licenses
- Biometric data
- Date of birth
- Place of birth
- Business phone numbers
- Employment information
- Credit card numbers
- Financial account numbers
- Medical records

To further investigate the system's adherence to PII protection, an assessor will need to review the following documents:

- Privacy impact assessments for policies, processes, and results
- Privacy threshold analyses, including policies, methods, and results

- System of Records Notices (SORN), when appropriate
- System security or privacy plan
- Security or privacy policy
- Privacy policy (for web applications)

Interviewing for PT — Personally Identifiable Information (PII) Controls

The assessor should interview someone knowledgeable about PII, the PII protection procedure, privacy assessments, and security compliance. This role is typically held by an information system security officer, compliance officer, privacy officer, cybersecurity expert, or system administrator familiar with privacy regulations. Key questions include:

Policy and Procedures:

- What is the purpose of PII processing?
- How is consent obtained for PII collection?
- Is there a System of Records Notice (SORN)?

Technical PT questions:

- Does the system's web page have a privacy statement?
- Is the privacy statement visible to users?

- How does the web page obtain consent to use PII?
- Does the website use cookies to gather user information?
- How is PII protected on the system?
- Does the company maintain records containing PII?

Operational and Management PT questions:

- Does the organization's policy include PII and privacy?
- Does the system gather PII?
- Has the system undergone a Privacy Impact Assessment (PIA)?
- What is the purpose of processing, storing, and transmitting PII on the assessment system?
- How is PII secured on the assessment system?

Testing PT — Personally Identifiable Information (PII) Controls

If the system under review contains a web page or content management system that collects or records PII, the assessor must examine the system's Behavior for the following:

Privacy Policy

The assessor should ensure the privacy policy is accessible on the website. This may appear as a pop-up or in the page's header, body, or footer. The policy should inform users about how the organization collects and utilizes information and should be consistent with the organization's privacy policy.

Consent

The assessor can verify consent by accessing the system under assessment. Users should be informed that their information will be gathered and monitored if they proceed. This is commonly a pop-up on the main page with an option to accept or decline.

Privacy Transparency

For federal systems that gather public information, users must be notified about the organization's use of their PII and its protection. The assessor will learn about this on the system's primary website.

Testing pop-ups and permission is only essential for systems with websites that gather PII. These settings let the user know if the system is processing, storing, or sending their PII. PT controls involve informing users about the purpose and nature of data collection and obtaining their consent.

Red Flags of Access Management

Some institutions request unnecessary personal information, such as social security numbers and dates of birth. The military is known for this. The issue arises when they attempt to secure data after collecting all that PII.

Green Flags of Access Management

When a business utilizes as little PII as feasible, it demonstrates foresight. They understand the importance of protecting personal data, therefore they only use a portion of the social security number when necessary.

Chapter 23

RA — Risk Assessment

Is the organization continuously monitoring vulnerabilities, running network scans, and documenting identified risks? SCAs look for this while researching RA control families. The company should have a policy detailing how frequently scans are performed on the assessed system. The RA control family encompasses all components of risk assessment, including security classification, risk response, privacy impact assessments, and threat hunting, in addition to vulnerability scanners. The most commonly used and visible RA controls include:

- **RA-1**, Policy and Procedures
- **RA-2**, Security Categorization
- **RA-3**, Risk Assessment
- **RA-5**, Vulnerability Monitoring and Scanning
- **RA-7**, Risk Response

Examine the RA — Risk Assessment Controls

The assessor will review the security policy to determine if the business has documented the necessity for frequent risk assessments, their frequency, and scope.

Once the assessor understands the expected outcomes, they may evaluate the current situation. The following documents provide evidence of risk assessments:

- Current risk assessment reports
- System security plans
- Scheduled vulnerability scans
- Privacy impact assessment reports (if applicable)

The system security strategy will indicate whether evaluations are being conducted. A well-maintained system security strategy includes up-to-date implementation statements, papers, modification dates within the past year, and proof of ongoing control monitoring. The analyzed system should have a security classification, often categorizing threats as Low, Moderate, or High.

When evaluating RA controls, assessors consider whether evaluations are conducted internally or externally. For SC and SI controls, the accuracy of data is more essential than the installation of security features.

Interviewing for RA — Risk Assessment Controls

RA controls primarily focus on the process rather than the tools. Interviews are an effective technique to examine an organization's process and ensure key personnel understand their responsibilities.

If the SCA has reviewed the policy, they will have an understanding of the organization's risk assessment process. If an interview is still necessary, it should be sufficient to confirm compliance with documents.

Here are some questions to assess if the RA process fits the RA controls:

- When was the last risk assessment done?
- What is the current security classification for systems?
- Which vulnerability scanning tool is used?
- How often do network scans occur?
- Who reviews the scan results?
- What happens if vulnerabilities are discovered?
- Who conducts network and vulnerability scans?
- Who performs the privacy impact assessment?
- Is there a security operations center?
- Is threat hunting conducted?

Testing RA — Risk Assessment Controls

The SCA can test an organization's evaluation methods. Assessors often assign a subject matter expert to execute an activity on the system and observe the results. Risk assessments may be performed by various systems, such as content management systems, scanners for networks, vulnerabilities, security compliance, and wireless connectivity, and software for conducting phishing campaigns.

Risk assessment tools should not be mistaken for the actual process or activity. Risk evaluations encompass more than just running a Nessus scanner. They also involve understanding the assets, assessing their weaknesses, and detecting potential threats.

Red Flags of RA Controls

RA includes risk assessments, vulnerability monitoring, risk response, and data analysis. Most organizations conduct risk assessments and self-analyses. Problems arise when risk assessments are not conducted regularly.

Green Flags of RA Controls

A business with diverse risk assessment methods is exceptional! They conduct web application scanning and penetration tests on their external and DMZ networks, perform routine internal scans, and combine physical and policy inspections. An effective risk assessment program integrates various evaluation methods to ensure comprehensive coverage and security.

Chapter 24

SA — System and Services Acquisition

Assessors should divide SA controls into three areas: acquisitions, systems, and software engineering. They evaluate the organization's equipment purchases and funding for acquisitions, as well as software development and system engineering practices.

Examination of the SA Controls

The assessor must examine three areas to get a comprehensive understanding of the SA controls:

- System engineering documents
- System acquisitions and services
- Software development

Not all of these will apply to every system being assessed. For example, if there is no software development on the assessed system, this part of the SA control family will not apply.

Examination of System Engineering Documents

In the case of system engineering, the assessor will consider whether the system has undergone a formal development process. A formal process often generates engineering documentation, meeting minutes, and diagrams. A successful system development approach will integrate security considerations early on. Even if the systems have been in existence for years, there should be recent engineering documentation and meetings discussing improvements and updates. The evaluation relies heavily on security measures such as SA-3, which follows the typical system engineering development life cycle:

- Adhere to the Security and Privacy Engineering Principle (SA-8) and implement security controls throughout the engineering process.
- SA-5, System Documentation, creates and maintains engineering documentation for the system.

Examination of Software Engineering

For systems with in-house development, the assessor will evaluate the organization's software engineering process. Significant controls include:

- **SA-10**, Developer Configuration Management: Ensures code and systems undergo configuration management procedures.
- **SA-11**, Developer Testing and Evaluation: Guarantees the implementation of security and privacy protections in the code.

The software libraries used to write the code must be safeguarded. Critical files should only be accessible to those who require them. Major software modifications require a configuration management procedure involving all stakeholders, including an information security representative and system owner. Document major software changes and meetings to ensure compliance with security controls, such as SA-10. Conduct testing to evaluate code before putting it into production. Examine the software engineering process by looking at the following:

- Software change configurations meeting notes
- Software development policy
- Software development process
- Results of software development tests
- Evidence of a secure software library

The most used and visible SA controls include:

- SA-1, Policy and Procedures
- SA-3, System Development Life Cycle
- SA-4, Acquisition Process
- SA-10, Developer Configuration Management

Examination of System Acquisition and Services

The assessor evaluates if the company purchases equipment and services safely. To acquire equipment and services for the evaluated system, the organization should adhere to the acquisition procedure (SA-4). The studied documents may not always be included in the information security documentation (e.g., a phrase in the security policy). Some firms have a distinct department that handles contracts, agreements, and acquisitions. The assessor can check the following documents:

- Contracts for information system assessment
- Acquisition policies and procedures
- Organizational agreements

The assessor is looking to determine if the organization has an acquisition procedure and if existing contracts for the evaluated system may be extended. They also want to ensure that the company does not purchase mission-critical services and parts from untrustworthy vendors.

Interviews on SA — System and Services Acquisition

To assess software development, the assessor should consult with information system security professionals knowledgeable in software engineering methods and safeguards. Software engineers and managers might be valuable resources. Questions include:

- Does software development involve change management?
- Does software development have a configuration management procedure?
- What is the software development process?
- How are software libraries protected?
- Is there a policy for software engineering?

Interviewing for the System Engineering Process

When interviewing for the system engineering process, managers, information security officers, and system engineers can provide valuable insights about SA controls. The questions address the organization's engineering procedures and ensure that security is included in the system engineering process:

- Did the system follow a system engineering process?
- Do modern systems include built-in security and privacy controls?
- How are security and privacy measures evaluated during large system changes? Are they mentioned at configuration management meetings?
- Are there any system engineering documents?

Interviewing for the Acquisition Process

Information system security officials, compliance specialists, and managers will understand the acquisition process. If not, they will be able to contact the appropriate person. If the organization has an acquisitions office, contact these individuals. Questions include:

- What is the purchase process?
- How does the organization obtain new system parts?
- Are there any system services? Examples include cloud computing, cybersecurity, internet connection, and system-critical services.
- Is the maintenance contract up-to-date?

Red Flags of SA Controls

When developing software for multiple systems, it's common to overlook the associated safety measures. Software developers may not understand or prioritize security controls and configuration processes. Assessors should study this area if appropriate.

Green Flags of SA Controls

Managers who have previously worked as system or software engineers understand the need to implement security early in the process. If the assessor can arrange an interview, these individuals can be valuable resources.

Chapter 25

SC - System and Communications Protection

When most people think of evaluations, they typically think of the assessment of SC controls. The "System and Communications Protection" (SC) family includes network, endpoint, and server scanning. However, SC control evaluations extend beyond simple scanning; they also involve reviewing policies, methods, procedures, and audit logs. To find vulnerabilities, assessors might need to test network access from the DMZ, the outside world, and internally.

A vast array of SC controls covers the organization's enterprise, and the assessor must stay within the parameters of the security assessment plan. It's important to focus on target IP addresses or hostnames to avoid introducing vulnerabilities not covered by the plan.

Organization of SC Controls

As of this writing, there are more than fifty SC controls. The SCA (Security Control Assessor) focuses on the objects being examined to determine whether the system is being safeguarded, rather than concentrating on every control and control enhancement, which might not apply to the system being evaluated. One effective way to organize the evaluation of SC controls is by categorizing the impacted assets:

Servers: Multiuser devices with centralized functionality. The assessor searches for vulnerabilities and flaws in the server's functioning.

***Endpoint Devices*:** Laptops, desktop computers, notepads, and other gadgets used regularly. If not configured properly, these can be the main point of entry for attackers.

***Network*:** Comprises interenctworking devices (switches and routers). SC controls focus on the system's border, created by the network's configuration.

Personnel: Individuals who administer and utilize the network, its auxiliary equipment, and the artifacts that document and map it.

Examination of SC Controls

Examination of SC Controls for the Network

The assessor utilizes a combination of testing, observation, and interviews for all control families of the SC controls. The focus will be on the examination approach, reviewing testing-derived vulnerability assessment reports, and possibly asking questions while observing the outcomes. Reviewing previous assessment results can provide insight into recurring issues and their context.

The network documentation should include recommended security measures such as:

- Encrypting important information
- Updating network software (antimalware software, network devices)
- Imposing access restrictions through security guidelines
- Collecting event logs
- Revising network documentation
- Teaching network users
- Implementing multiple protection layers
- Managing network ports, protocols, and services

The assessor will examine documents like network diagrams, previous network scans, network engineering documents, security guidelines, lists of ports, protocols, and services, system security plans, privacy impact assessments, encryption engineering papers, networking procedures, and network usage security training.

Examination of SC Controls for Server and Endpoint Devices

The SC assessment will involve a deep dive into the settings and communication occurring on servers and endpoint devices within the environment. Assessments will cover all types of servers, from virtual cloud-based images to on-premises big data clusters. The assessor will observe how the company handles the security characteristics of each system.

***Key controls include*:**

- ***SC-2:*** *Separation of System and User Functionality*: Ensuring that regular users cannot modify system settings controlled by privileged users as needed.
- ***SC-4:*** *Information in Shared System Resources*: Using role-based group creation, permission-based grouping, and data encryption to prevent unauthorized information transfer.

The assessor reviews processes, documentation, and scan findings, looking for configuration management, auditing, identity and authentication, access controls, and security best practices. If gaps are found, further investigation is required by reviewing additional documents, asking questions, and conducting tests.

Additional best security practices include:

- **Least functionality (CM-7):** Systems are constructed with only necessary features.
- **Least privileges (AC-6):** Ensuring users have the minimal privileges necessary.

- **Baseline Configuration (CM-2):** Ensuring system configurations align with organizational standards.
- **System backups:** Ensuring critical data is backed up as required.
- **Vulnerability management (SI-2):** Setting deadlines for fixing system vulnerabilities.
- **System support (SA-22):** Managing or removing unsupported applications and replacing end-of-life systems.
- **Change controls (CM-3):** Using impact analyses, meetings, and documentation to manage system changes.

The assessor reviews documentation such as baseline configuration guidelines, Group Policy Object (GPO) standards, security guidelines for endpoint and server configurations, access control documentation, and audit logs showing endpoint and server access restrictions.

Interviews of SC Controls

Network and Network Components

The assessor should consult network operators, network engineers, or IT professionals familiar with the network's principles and structure. Key questions include:

- Do network policies and procedures exist?
- Is network segmentation required?

- Does the network have DMZ, external, and internal components, and protection against denial of service?
- Are there network shares, and what safeguards are in place?
- Is there network surveillance, and how is it conducted?

Servers and Endpoint Devices

The best points of contact are system administrators. Specific questions include:

- Is there a policy or process for network configuration?
- What roles and groups exist on the endpoint devices and servers?
- Do typical users have local administrator privileges?
- Are there shared folders, and are they locked down by user roles?
- Are host-based safeguards present?
- Are conversations on servers and endpoints observed?
- Does the company separate systems necessary for its mission?
- How is the system designed to protect personally identifiable information?
- How are systems with classified data safeguarded?

Testing of SC Controls

The SCA must be familiar with scanning tools and procedures to test an organization's network. Before proceeding, written consent specifying the scope, type of testing, timing, and personnel administering the test is required.

- Types of Network Scans
- Internal network scans
- System compliance scans
- Security policy platforms
- External network scans
- DMZ scans
- Penetration testing
- Web application scans
- Wireless scans
- Firewall scanners

Credentialed vs. Non-Credentialed Scans

Credentialed Scans: Require identification and authentication (e.g., username and password). Used for compliance, endpoint, DMZ, and internal scans.

Non-Credentialed Scans: Do not require authentication. Used to identify what is visible to attackers and unauthorized users, typically for penetration tests and external scans.

The Scan Results

Popular scanning tools include Qualys, Rapid7, and Tenable. These tools assign a severity level to vulnerabilities, using the Common Vulnerabilities and Exposures (CVE) mechanism and the Common Vulnerability Scoring System (CVSS). Severity levels range from 0.0 (none) to 10.0 (critical).

The assessor documents scan results, leaving it to stakeholders and system owners to address the identified risks. Other scoring systems include the Common Weakness Scoring System (CWSS), primarily used for software code development.

Interpretation and Accuracy

It is the assessor's responsibility to ensure the scanner functions properly. Credentialed scans may fail if authentication fails for target systems. Ensuring the correct range of systems is scanned and having a knowledgeable individual to handle technical issues is crucial.

- Red Flags of SC Controls
- Many critical results from scans of Internet-facing and end-of-life systems in the DMZ.
- Lack of a network diagram.
- Outdated network information.

- Network documentation that does not match the actual network.
- Public IP addresses on the internal network.
- Publicly accessible high-value systems.
- Sensitive information networks that are not segregated or protected.
- No devices, such as firewalls, between networks with different security postures.

Green Flags of SC Controls

Conducting penetration tests and thorough internal and external network scans indicates a serious approach to network security.

System and Communications Protection Resources

- **NIST SP 800-189**, Resilient Interdomain Traffic Exchange
- **NIST SP 800-77**, Guide to IPsec VPNs
- **FIPS 186**, Digital Signature Standards (DSS)
- **NIST SP 800-63-3**, Digital Identity Guidelines
- **FIPS 140-3**, Security Requirements for Cryptographic Modules
- **NIST IR 7788**, Security Risk Analysis of Enterprise Networks Using Probabilistic Attack Graphs
- **NIST SP 800-41**, Guidelines on Firewalls and Firewall Policy
- **NIST SP 800-56**, Recommendation for Key-Derivation Methods in Key-Establishment Scheme

- **NISTIR 7956**, Cryptographic Key Management Issues & Challenges in Cloud Services
- **NISTIR 7966**, Security of Interactive and Automated Access Management Using SSH
- **NIST SP 800-57 (Parts 1-5)**, Recommendation for Key Management

Chapter 26

SI - System and Information Integrity

Information integrity and system security are among my responsibilities. Within this control family, the assessor verifies the effectiveness of vulnerability management, spam and antivirus software, system monitoring, information input validation, and other measures that safeguard the organization's data from unwanted access or alteration. The primary controls that form the foundation of the SI family include:

- **SE**, Policy, and Procedures
- **SI-2**, Flaw Remediation
- **SI-3**, Malicious Code Protection
- **SI-4**, System Monitoring
- **SI-10**, Input Validation (for web applications)

These SI controls provide the assessor with valuable information about the overall risk to the assessed system. Numerous variables, including the environment, mitigating measures, and system risks, can affect this.

Examination of SI - System and Information Integrity Controls

To determine how the organization addresses input validation, malicious code protection, system monitoring, and flaw remediation, the assessor should review the relevant policies. These controls cover various aspects such as web pages that collect user input, antivirus software, audit log monitoring, and vulnerability management.

Flaw Remediation (SI-2)

Flaw remediation involves fixing systems once bugs, upgrades, or new vulnerabilities are found. Key items for examination include:

- The strategy and plan for vulnerability management
- Process results of the network scan
- A strategy for remediation
- A schedule and benchmark for updating software
- Documentation for patch management
- Outcomes of patch management
- Security plan for the system
- Policy for system security

The assessor looks for an efficient method within the organization to quickly address any issues.

Antimalware, System Monitoring, and SPAM

The SI family includes antivirus, antimalware, endpoint protection response, and antispam software. The assessor must determine if the organization has chosen a malicious code protection (SI-3) solution and whether it has been deployed on the network or installed on the target system. Some organizations fail to frequently update their virus signatures, which can be problematic.

The organization may need to manually update the signatures or manage them centrally if some systems cannot connect to the vendor for automatic updates. The organization or system being evaluated may not require antispam (SI-8) if email is not processed by the system. If it does, the assessor will check if the antispam solution has been installed and activated on the system.

Antimalware programs are often part of more comprehensive system monitoring programs. The assessor will look for a procedure wherein the company monitors audit records for behavioral patterns that might indicate an intrusion or unauthorized activity (SI-4). This could be part of an information and event management system, a cybersecurity analyst team, or a security operations center. For small networks or stand-alone systems, the monitoring procedure may involve a quarterly event log review. The assessor will search for documentation of these actions.

Items the assessor can examine include:

- System security policy and strategy
- A software inventory that includes antivirus programs
- The system baseline showing the presence of antivirus software
- System scan results indicating outdated antivirus software
- Screenshots showing the installed antimalware program
- Incident response records
- Protocols for responding to incidents

Examination of Information Input Validation

For systems with frontend interfaces, including web apps or databases that gather user data, the assessor may review information input limitations. User input should be verified by the system to ensure proper data entry. For example, a data input form may require a user's first name to contain only letters. Input fields accessing databases may have character restrictions to prevent attackers from inputting complex commands.

The assessor should look for documentation outlining permitted inputs, such as standards, plans, and procedures. Additionally, the organization may provide messages and metadata informing users of the expected input. Other tools for observing information input validation applications include:

- Configuration settings of the software input box

- Application policy
- Work instructions detailing user input restrictions
- User instructions with input restrictions
- Source code

Interview for the SI Family

The assessor's focus on SI controls may require interviewing various stakeholders. Cybersecurity experts will be knowledgeable about antivirus software and system monitoring. While spam and email-related issues often have their teams, cybersecurity should be aware of the security aspects of content filtering within the organization.

Input validation is typically performed by software developers or system administrators. These roles vary across organizations, so interviews need to be scheduled ahead of time to ensure the necessary staff is available for evaluation.

Insightful questions to ask include:

- Is there a policy addressing patching, antivirus, monitoring, or information integrity? (SI-1)
- What is the vulnerability management process for this system? (SI-2)
- How are vulnerabilities identified, reported, and corrected? (SI-2)
- How are systems tested before being patched or upgraded? (SI-2)
- Is malicious code protection (antivirus) used on the system? (SI-3)
- How are virus signatures updated for the antivirus software? (SI-3)

- How does the antivirus system work? Is it centralized? (SI-3)
- Is there a method to detect potential attacks on the network? (SI-4)
- How does the organization identify unauthorized uses of systems? (SI-4)
- Are potential security incidents and anomalies analyzed? (SI-4)

Testing for the SI Family

The assessor must examine how SI tools, applications, and procedures safeguard data and systems integrity. Subject matter experts may need privileged access to certain systems, likely the same individuals required for the interview.

The assessor should be familiar enough with the tools and applications being examined to understand their functionality. Testing instruments include:

- Antimalware/Antivirus (SI-3, SI-4)
- Endpoint protection systems (SI-3, SI-4)
- Spam filters
- Endpoint detection and response systems (SI-3, SI-4)
- Security information and event management systems (SI-3, SI-4)
- Intrusion detection systems (SI-3, SI-4)

These are just a few available tools for testing. When using complex tools, it is advisable to allow a cybersecurity expert or system administrator to log in and operate the system. The assessor should ensure that the instruments used are appropriate for the organization and current. Does the solution adhere to the organization's security guidelines?

Red Flags of SI Controls

A warning sign is when a phishing attack compromises the organization's network through security events. Malware downloaded from the Internet can exploit software bugs. The assessor will identify several issues if the organization is not fixing the network's systems.

A common problem is a lack of vulnerability management, which can reveal:

- Many end-of-life systems in mission-critical parts of the system
- Overwhelming numbers of vulnerabilities
- No plans to fix the vulnerabilities

Green Flags of SI Controls

A top-notch organization will have a strong vulnerability management program that addresses vulnerabilities at a rate commensurate with the severity of the findings.

SI - System and Information Integrity

Resources

- NIST SP 800-128, Guide for Security-Focused Configuration Management of Information Systems
- FIPS 186, Digital Signature Standards (DSS)
- FIPS 140-3, Security Requirements for Cryptographic Modules
- NIST IR 7788, Security Risk Analysis of Enterprise Networks Using Probabilistic Attack Graphs
- NIST SP 800-40, Guide to Enterprise Patch Management Technologies

Chapter 27

SR - Supply Chain Risk Management

To ensure that all resources—including tools, materials, and services—are sourced securely and not compromised, supply chain risk management (SR) controls are evaluated by the organization's procedures. For example, if the company is buying external hard drives from a vendor, it must ensure the vendor's operations, offerings, and customer service are reliable and authorized. If the vendor is subject to any federal, state, or industry limitations, the organization should be aware of them. The assessor will look for upper-level management's authorization and documentation of this approach, and there may even be a dedicated job function for supply chain risk management.

Supply chain risk management is everyone's responsibility, even if a specific position or department, like acquisitions, is devoted to ensuring that the goods and services used by the company are authorized and secure. If the assessor observes, for example, the chief financial officer providing the office with flash drives purchased from Walmart using company funds, bypassing all formalities, and then plugging them into the work systems because it's "no big deal," they should include this information in their risk assessment report.

Among the most prominent SR controls are:

- **SR-2**, Supply Chain Risk Management Plan
- **SR-3**, Supply Chain Controls and Processes

Examination of SR - Supply Chain Risk Management Controls

There is no "one size fits all" approach to supply chain risk management (SCRM) strategies. Each firm has unique objectives and business demands, which determine its supply requirements and legal restrictions. When evaluating the supply chain, the assessor must review the organization's Supply Chain Responsibility Management (SCRM) to understand the activities it claims to be performing. The assessor should look for procedures and policies that are current, authorized in line with the organization's guidelines, and compliant with industry standards.

The organization's goals and business needs determine what needs to be included in the SCRM and what needs to be delivered. As assessors, we cannot dictate what needs to be provided.

The assessor needs to determine if the company considers supplier risks. For instance, the SCRM policy could discuss the significance of the vendor's cybersecurity posture and supplier hazards if the tools or services require a direct network connection. To remain a supplier, the vendor might need a recurring evaluation or certification.

Another example would be that the company is restricted to purchasing software from a verified, approved application list. The assessor focuses on the organization's supply chain process, the risks associated with the goods and services provided, and the source of those supplies. This should be approved by a document management procedure, strategy, policy, or other framework.

The following documents may be reviewed by the assessor:

- System security plan SR implementation statements
- Contract or agreement language mentioning governance, risk, or compliance
- SCRM plan, policy, or procedures

- Acquisition documents with purchasing restrictions or risk management strategies
- Business-to-business network interconnection agreements
- Shipping and trade policies
- Export controls policy
- Standards that match international regulations on trade

Interviews for SR - Supply Chain Risk Management Controls

Professionals in acquisitions, HR, cybersecurity, management, and compliance will be able to guide you toward the best supply chain risk management strategies. This may be a separate unit within the business or integrated into acquisition, compliance, or HR activities.

As SCAs, our responsibility is to determine if the company has an SCRM process, how well it is documented, and how it complies with the pertinent NIST 800 SR controls for the system under assessment. During the interview, it is possible to gauge the subject matter experts' knowledge of the SCRM and their actions in compliance with their policy.

Insightful questions to ask include:

- Are the agreements and contracts with outside vendors current?
- Are the contracts with outside providers still in effect?
- Do agreements and contracts cover pertinent compliance issues (e.g., PCI compliance in retail contracts or federal systems requiring specific vendor support)?
- Do agreements and contracts mention compliance?
- Is there a mechanism for managing supply chain risks?
- Is there a record of the supply chain risk management procedure?
- Are there limitations on the goods or services that can be purchased (e.g., a list of acceptable manufacturers or software/hardware combinations)?
- Have approved sources been used to get hardware and software providers?
- Are maintenance contracts current?
- Is there a policy for exporting goods and services internationally (if relevant)?

Testing for SR - Supply Chain Risk Management Controls

The company may have contracting and acquisition tools to manage supply chain risk, in addition to documents (SR-5).

These tools may be tested by the assessor to confirm how well they support the organization's risk management efforts. These instruments could include:

- Monitoring the status of shipments – shipments may need to be made to and from secure sites, with discretion being an issue.
- Secure and controlled order process – only certain employees can place orders.
- Systems for managing suppliers.
- Security rating systems like BitSight, which rates suppliers' cybersecurity according to risk.
- Systems for vendor compliance.

The assessor will need a subject matter expert to guide them through and explain the functions of each SCRM instrument. As the expert logs in and uses the system, the assessor can ask relevant questions. For example, "Does the organization use BitSight to determine the risk of vendors they work with?" and "Can you show me the risk score for one of your vendors?" or "What is the organization's procedure for utilizing vendor compliance systems?"

BitSight is just one example; most enterprises won't use all the technologies listed here. The goal is to prompt the system administrators as they navigate the system to determine who is being utilized to manage the risk.

Red Flags of SR Controls

If staff members use their own money to purchase essential components from unidentified vendors, this should raise many concerns. As the assessor, you should find out if there is a procedure they should be following and make the staff members aware of it.

Green Flags of SR Controls

For medium-sized and larger businesses, automated supply chain tracking and control systems are a great sign. The organization has done its due diligence when acquisitions, IT, and cybersecurity are aware of the supply chain procedure.

SR - Supply Chain Risk Management Resources

- **NIST SP 800-128**, Guide for Security-Focused Configuration Management of Information Systems
- **FIPS 186**, Digital Signature Standards (DSS)

- **FIPS 140-3**, Security Requirements for Cryptographic Modules
- **NIST IR 7788**, Security Risk Analysis of Enterprise Networks Using Probabilistic Attack Graphs
- **NIST SP 800-40**, Guide to Enterprise Patch Management Technologies

Made in the USA
Coppell, TX
29 October 2025

62071113R00128